NTERNATIONAL SUB-CONTRACTING: A TOOL OF TECHNOLOGY TRANSFER

sian Productivity Organization

OKYO

INTERNATIONAL SUB-CONTRACTING: A TOOL OF TECHNOLOGY TRANSFER

1978
Asian Productivity Organization
TOKYO

Designed and Printed in Hong Kong by
NORDICA INTERNATIONAL LIMITED
for
ASIAN PRODUCTIVITY ORGANIZATION
4-14 Akasaka 8-chome,
Minato-ku, Tokyo 107, Japan

ISBN: 92-833-1047-0 Casebound
92-833-1048-9 Limpbound

Forword

The sub-contracting system has, of late, achieved considerable importance in various developing countries. They have taken a cue in this regard from Japanese experience where small industries have played and are still playing an important role as sub-contractors to their parent firms.

Under the system of sub-contracting,a vertical relationship is formed between two parties which develop a complementary, not necessarily supplementary, linkage in the economic development. Under the umbrella of the parent firms that provide necessary assistance relating to technology, marketing and management, the sub-contracting firms in return make a dent in the production plan where parent firms may find it uneconomical to deal directly with the whole spectrum of production.

Sufficient information has now become available with the passage of time as far as this system works intra-nationally. But very little information or documented material is available on international sub-contracting though increasing recognition is given to its role in this age of global business. Needless to say that international sub-contracting results in accelerating employment opportunities, entrepreneurial development, market expansion and export promotion and in increasing foreign exchange earnings and above all improvement of skills.

It is against this background that the Asian Productivity Organization (APO) decided to carry out a fact-finding survey on "international sub-contracting and transfer of technology and skills" in some of its selected member countries in 1975-76 in collaboration with the International Labour Office (ILO) that also showed an interest in this subject. The survey was intended, *inter alia,* to collect and compile information on the extent of practice of international sub-contracting in some of the APO member countries and assess its current and potential impacts on the technological standard and productivity of their industries. It also attempted to identify major difficulties and problems involved in international sub-contracting and transfer of technology and skills from parent firms.

The ILO provided the services of Mr. Susumu Watanabe, Economist, Technology and Employment Branch, Employment and Development

Department, International Labour Office, who, acting as the Chief Expert, provided a common framework and guidance for undertaking the survey by the local experts in the APO member countries of Hong Kong, India, Republic of Korea, and Pakistan. The Chief Expert himself collected information in Japan, the Philippines and Singapore.

Mr. Watanabe finished his survey report in 1976 and the APO decided to hold a study meeting which was held in Tokyo from 16th to 19th November 1976. It was attended by five experts from APO member countries, namely Hong Kong, India, Pakistan, the Philippines and Singapore, three experts from the Council of Asian Manpower Studies and an observer each from the Institute of Developing Economics, Tokyo and Ministry of Labour, Government of Japan. (For list of participants see Appendix I.)

Besides examining the survey report which dealt with the spread of international sub-contracting, its determinants and problems, the Study Meeting heard country reports from the survey countries and report on a comparative study on ancillary firm development in Asian countries currently carried out by the Council of Asian Manpower Studies. Representatives of two Japanese corporations, M/s. Ishikawajima Harima Heavy Industries Complex (IHI) and Matsushita Electric Industrial Company, explained their experience relating to international sub-contracting and transfer of technology and skills to developing countries. Reports from the participants and discussion on the survey report brought to the surface various aspects of the subject which were found lacking in the survey. The Study Meeting thus made a number of important recommendations. (For detailed recommendations, see Appendix II.)

The present publication, which is in two parts, contains the survey reports in Part I and country reports with well-referred supplementary data in Part II. The APO hopes that this publication, which is meant to be of a pioneering nature, would be both interesting and informative to readers who have aspired to laying their hands on such a study.

Taking this opportunity, the APO wishes to record its sincere thanks to the ILO, National Productivity Organisations of member countries, local experts engaged in the survey and the Council for Asian Manpower Studies for their valued cooperation. Special thanks are due to Mr. Watanabe for his untiring efforts in making it possible for the APO to undertake such a project of pioneering nature and for the time spent by him to edit the country reports despite of his official preoccupation at the ILO.

Asian Productivity Organization

Tokyo
February 1978

Preface

A lot of lip service has been given recently to international sub-contracting, which is considered, not without good reasons, to have contributed to the development of the Third World Economy, e.g. by accelerating industrialisation and creating employment and income. Very little is known, however, about the actual state of this business practice, particularly where developing countries are involved. This survey may be said to be one of the first attempts to fill such an information gap.

Time and resources available for the task did not permit, however, such a scale of survey as would be needed for a systematic quantitative analysis even within the countries selected for the survey. The number of samples was sometimes extremely small: only 10 in the case of Hong Kong. Our coverage of parent firms was largely confined to Japanese firms, the survey area being limited within the APO region. In preparing the present report, however, efforts were made to enlarge the scope of the study as far as possible by using supplementary material obtained during my own field work. Still, the picture which has come out of our survey and which is presented here cannot but be partial. In the almost total absence of factual information on the subject, however, I still believe that the results of our survey will throw some useful light on certain aspects of international sub-contracting. At the same time it is hoped that shortcomings in our survey and reports will be made good by further empirical studies on this subject.

The survey was conducted in four developing countries (Hong Kong, India, the Republic of Korea, and Pakistan) between the summer of 1975 and the spring of 1976. In each country, a local expert or group of experts was recruited for this purpose through the National Productivity Organisations, according to the customary procedure of the APO. I was asked to prepare a common framework of the survey and co-ordinate the country surveys, as well as to write a main report on the basis of their findings and any additional material I could obtain. To strengthen our data base, I undertook a questionnaire survey in Japan and visited the Philippines and Singapore for several days each.

The main report and the four country reports are presented in Part I and Part II of this publication, respectively. All the experts used the same terminology that is explained in the introductory chapter of the main report together with the objectives of the survey.

I am grateful to many companies, institutes and individuals in various countries for their cooperation in this survey. I am particularly grateful to the APO, the Hong Kong Productivity Centre, Korea Productivity Center, the National Productivity Board and the Economic Development Board of Singapore, the Development Academy of the Philippines, and the Department of Science and Technology of India, for their generous help in arranging my visits. The Tokyo and Japan Chambers of Commerce and Industry kindly supported me in conducting my questionnaire survey in Japan by printing and distributing the questionnaire forms among their members and collecting replies from them. The United States Mission in Geneva supplied me with valuable data. I would also thank Professor Shigeru Ishikawa and my colleague Agit. S. Bhalla for their constructive comments on an earlier version of this report. I am also grateful to the APO for partly financing my work by the APO's Oshikawa Fellowship for the year 1975.

The views expressed in this report are my own and they are not necessarily shared by the APO and ILO.

Susumu Watanabe
Technology and Employment Branch
Employment and Development Department
International Labour Office

Geneva
October 1977

Table of Contents

PART I: SURVEY REPORT

Chapter I

Scope and Objective of the Survey

INTRODUCTION

One of the most remarkable developments in the domain of international economic co-operation in recent years is the spread of sub-contracting arrangements between developed and developing countries. This practice merits further encouragement for at least two reasons. First, it will contribute immediately to the well-being of all the nations involved by distributing income and job opportunities over a wider area than otherwise and by enhancing the efficiency in the utilisation of resources. Second, in the long run, it will help foster strong independent local industries out of sub-contractors as a result of various kinds of assistance provided by the parent firms, and accumulation of business experience which is essential for industrialisation and yet badly lacking in the developing countries.

The first point seems quite obvious and probably needs no further elaboration. As regards the second statement, the history of Japanese modern industries is full of supporting evidence. The Matsushita Electric Industrial Company has grown into one of the world's largest companies from a petty sub-contractor who manufactured insulators.[1] Ishikawajima Harima, Mitsubishi, and some other giant firms have been enthusiastic about receiving sub-contracts from the Rolls-Royce, the General Electric etc. to manufacture parts and components of aircraft even at the sacrifice of short-term profits, largely in the hope of acquiring advanced technology from the later companies and gaining experience.[2]

Potential merits of international sub-contracting seem particularly great in the case of developing countries, where there are still very few indigenous firms which are financially and technically strong enough to help sub-contractors.

---

[1] Konosuke Matsushita, *Watakushi no Rirekisho,* My Personal History, Vol. 1 (Tokyo, Nihon Keizai Newspaper Co., 1957), pp. 247–248.

[2] *Nihon Keizai Newspaper* (Tokyo), 16 March 1973 (morning edition), p.8.

It was, therefore, opportune that the Asian Productivity Organization undertook a survey on international sub-contracting in its member states in collaboration with the International Labour Office.

OBJECTIVES OF THE SURVEY

The primary objective of the survey was to formulate policies for further promotion of international sub-contracting and transfer of technology and skills to the APO member countries through this business arrangement. In more concrete terms, it was attempted firstly, to collect and compile information and data on the prevalence of international sub-contracting in some of the APO member countries, secondly, to assess its current and potential impacts on skills and productivity in their industries, and thirdly, to identify major difficulties and problems involved in international sub-contracting and transfer of technology and skills from parent firms.

SCOPE AND ORGANISATION OF THE SURVEY

A country survey was conducted by a local expert in Hong Kong, India, Korea and Pakistan some time between autumn 1975 and summer 1976, using a common framework prepared by the chief expert and following his guidance. The chief expert visited these countries, except Pakistan which joined the survey later, for consultation with the local experts and for his own fact-finding tour in August and November-December 1975, when he also visited Japan, the Philippines and Singapore. In the last two countries, he confined his investigation to the Progressive Car Manufacturing Programme (PCMP) and the capital and training assistance schemes for supporting industries respectively, which are probably the most important attempts that have ever been made in this region for the purpose of encouraging international sub-contracting. Main findings in these two countries are summarised in the two annexes to this report. In Japan, some original information was collected by means of questionnaires, which were sent to 416 members of the Tokyo and Japan Chambers of Commerce and Industry, which belonged to the survey industries and which were located mostly but not necessarily in the Tokyo area.

Four industries were selected for the survey: textiles (mainly garment), electrical and electronic equipment, transport equipment, and precision equipment manufacturing. These are known to be the industries where sub-contracting has been most widely practised.[1] When an interesting case of international sub-contracting was discovered outside these industries,

---

1 Cf. Susumu Watanabe, "Sub-contracting, Industrialisation and Employment Creation", *International Labour Review*, July – August 1971, p. 55.

however, it was included in the survey. Printing industry in Pakistan and sports goods manufacturing in India and Pakistan are such cases.

DEFINITION OF TERMS

"Sub-contracting" is a business practice whereby the party offering the sub-contract (*parent firm* or company) requests another independent enterprise (*sub-contractor*) to undertake the whole or part of an order it has received instead of doing the work itself, while assuming full responsibility for the work vis-a-vis the customer. Sub-contracting differs from the mere purchase of ready-made parts and components "off the shelf" in that there is an actual contract between the two parties setting out the specifications of the order. The parent firm can be either a wholesaler or a retailer which does not undertake any manufacturing or processing of the product itself (e.g. a department store which has various goods manufactured to sell under its own brand), or it can be a manufacturer which uses the purchased product as parts or components of its own produce (e.g. a car manufacturer which sub-contracts the manufacturing of gear boxes to metal engineering firms). For subsequent discussion it is convenient to call these two types "commercial sub-contracting" and "industrial sub-contracting", respectively. Sometimes manufactures can become parent firms in commercial sub-contracting, a typical example being garment manufacturers which sub-contract their work to others, because of the shortage of labour.

Four types of international sub-contracting can be distinguished:

Cross-border commercial sub-contracting – Parent firm A (e.g. a Japanese shirt trader or manufacturer) receives orders for its products from customers at home (in Japan) or abroad (e.g. in the United States) and sub-contracts the entire manufacturing work to a firm in another country (e.g. a Korean garment manufacturer). The products are delivered to the customers either via firm A, or direct from the sub-contractor under firm A's brand.

Cross-border industrial sub-contracting – Parent firm A (e.g. an American T.V. maker) sub-contracts manufacturing or processing of parts and components to a firm abroad (e.g. in Singapore) for assembly at its own plant at home (in the United States) or in a third country (e.g. Hong Kong).

Strictly speaking, intra-firm transactions of parts and components between different branches of a multinational firm (i.e. between the *principal firm*[1] at home and its overseas subsidiaries, or between its subsidiaries abroad) is not sub-contracting, as they are not mutually independent firms, but this was included in the survey as a special case of cross-border industrial sub-contracting, partly because there is no way to isolate it statistically, except in our questionnaire survey in Tokyo.

1 Note that, in our reports, "parent firm" and "principal firm" are used as an antithesis of "sub-contractor" and "subsidiary", respectively.

Within-border industrial sub-contracting – Parent firm A is a foreign subsidiary (e.g. Ford, Philippines) which sub-contracts manufacturing of parts and components to local firms for assembly at its own plant. In this case, the sub-contracted parts and components do not leave the country of manufacturing at least before assembly, except when they are exported as replacement parts.

Within-border commercial sub-contracting – Parent firm A is a foreign subsidiary (e.g. Dunlop, India) which sub-contracts manufacturing of end products (e.g. tennis rackets) to local firms to distribute under its own brand.

Technology and skills to be transferred may be either of the three categories:

(a) product designs and specifications;
(b) production techniques and skills; and
(c) managerial techniques and skills, including quality control techniques and skills.

Chapter 2

Spread of International Sub-contracting

No systematic statistics exist to permit a comprehensive assessment of the spread of various types of international sub-contracting. There is, however, some evidence that this business practice has been expanding rapidly between certain developed and developing countries. Different patterns of sub-contracting seem to be concentrated in different countries.

CROSS-BORDER SUB-CONTRACTING

Sub-contracting from the United States

Item 807.00 of the United States Tariff Schedules provides that manfactured imports containing components of U.S. origin are subject to duty on their full value less the value of the U.S. fabricated components contained therein. Statistics of imports under this Item permit partial assessment of cross-border international sub-contracting from the United States.[1] The assessment based on these data is partial, because an important proportion of sub-contracting from the United States to other countries does not involve supply of materials. Even so, we can learn some interesting facts from these data which are summarised in Table 1.

First, during the 8-year period 1966–1974, international sub-contracting from the United States which involved supply of materials, more or less quintuplicated from 889.8 million dollars to 4,828.0 million dollars in terms of total value, and from 776.5 million dollars to 3,818.5 million dollars in net value (i.e. dutiable value). The growth rate was by far faster in the case of sub-contracting to developing countries than to developed countries: the former grew by 32 times (36 times in terms of dutiable value) while the latter increased by only 3.5 times (3.6 times). Consequently, the

[1] This provision is applicable to those foreign firms which buy U.S. fabricated components for assembly into final products exported to the United States. In 1969, some 60 per cent of U.S. imports under Item 807.00 were from such foreign firms (cf. J.M. Finger, "Tariff Provisions for Offshore Assembly and the Exports of Developing Countries," in *Economic Journal,* June 1975, p. 365).

Table 1

U.S. Imports under Tariff Item 807.00 by Group of Countries, 1966-74

(in millions of U.S. dollars)

Country	1966 Total Value	1966 Value of US contents	1966 Dutiable Value	1970 Total Value	1970 Value of US contents	1970 Dutiable Value	1974 Total Value	1974 Value of US contents	1974 Dutiable Value	Change over 1966-74 (1966 = 100) Total Value	Change over 1966-74 (1966 = 100) Value of US contents	Change over 1966-74 (1966 = 100) Dutiable Value
Grand	889.8(100.0)	113.3	776.5	2,008.7(100.0)	436.4	1,572.3	4,828.0(100.0)	1,009.4	3,818.5	543	891	492
Total	(100.0)	(12.7)	(87.3)	(100.0)	(21.7)	(78.3)	(100.0)	(20.9)	(79.1)			
Developed	829.0(93.2)	84.2	744.8	1,506.3(75.0)	161.2	1,345.1	2,899.9(60.1)	199.8	2,700.1	350	237	363
Countries	(100.0)	(10.2)	(89.8)	(100.0)	(110.7)	(89.3)	(100.0)	(6.9)	(93.1)			
Developing	60.5(6.8)	28.9	31.3	502.2(25.0)	275.2	226.9	1,927.6(39.9)	809.5	1,118.0	3,186	2,801	3,572
Countries	(100.0) (100.0)	(47.8)	(51.7)	(100.0)(100.0)	(54.8)	(45.2)	(100.0)(100.0)	(42.0)	(58.0)			
Mexico	7.0(11.6)	3.6	3.4	211.4(42.1)	133.6	77.9	917.6(47.6)	487.2	430.7	13,109	13,533	12,668
	(100.0)	(51.4)	(48.6)	(100.0)	(63.2)	(36.8)	(100.0)	(53.1)	(46.9)			
Hong Kong	41.4(68.4)	19.1	22.4	120.5(24.0)	64.1	56.3	200.1(10.4)	56.9	143.2	483	298	639
	(100.0)	(46.1)	(54.1)	(100.0)	(53.2)	(46.7)	(100.0)	(28.4)	(71.6)			
Korea	–	–	–	20.8(4.1)	14.8	6.0	66.8(3.5)	31.8	35.0	321*	215*	583*
				(100.0)	(71.2)	(28.8)	(100.0)	(47.6)	(52.4)			
Singapore	–	–	–	20.1(9.5)	11.2	8.9	137.4(7.1)	26.5	110.8	684*	237	1,245*
				(100.0)	(55.7)	(44.3)	(100.0)	(19.3)	(80.6)			
Taiwan	6.6(10.9)	3.3	3.3	85.9(17.1)	26.6	59.3	318.5(16.5)	59.8	258.7	4,826	1,812	7,839
	(100.0)	(50.0)	(50.0)	(100.0)	(31.0)	(69.0)	(100.0)	(18.8)	(81.2)			

Note: *Because of rounding, the figures may not add up to the totals shown. The marginal differences (0.5 million dollars or less) between the grand totals and the sums of the figures for the two groups of the countries are attributable to the centrally planned countries.*
– denotes a figure less than 0.5 million dollars.
The figures in the parentheses indicate the percentage composition of the grand totals and totals.
Those for the individual countries represent their shares in the developing world as a whole.
**1970 = 100.*

Source: *U.S. International Trade Commission: U.S. Imports for Consumption, Tariff Items 807.00 and 806.30, Specified Years 1966-74 (June 1975), supplemented by data from United States Tariff Commission: Economic factors affecting the use of items 809.00 and 806.30 of the tariff schedules of the United States: report to the President on investigation on No. 332-61 under Section 332 of the Tariff Act of 1930, (TC publication 339), (Washington D.C., September 1970), table 4.*

share of the developing world grew from a mere 6.8% in 1966 to 39.9% in 1974, in terms of total value.

Second, the benefit of the application of this Tariff Item vis-à-vis the developing countries has been concentrated in a small number of countries: Mexico, Hong Kong, Korea, Singapore, and Taiwan. In 1966, they accounted for over 90% of total imports from the developing world under this Traiff Item, and their share remained to be 85.1% in 1974. Meantime, however, the share of the Asian countries was lost substantially to Mexico, which accounted for 47.6% in 1974 as compared with 11.6% in 1966. In the case of Hong Kong, the share fell drastically from 68.4% in 1966 to 10.4% by 1974.

Third, the value added earned by the developing countries in international sub-contracting is, generally speaking, much lower than that earned by the developed countries: the ratio of dutiable value to the value of U.S. components is not much higher than 1.1 in the former case, while it is 9.1 or even higher in the latter case.

Fourth, it is, however, noteworthy that in all the Asian countries in the Table the weight of the U.S. components has been declining steadily and rapidly. Singapore and Taiwan have now almost reached the standard of the developed countries in this respect, although quite a proportion of the non-U.S. components must come from a third country.

As regards non-precious metal products, the U.S. Tariff Item 806.30 provides for a similar exemption from import duties on the condition that the re-imported article be processed further in the United States. Imports under this Item grew from 63.2 million dollars in 1966 to 544.5 million dollars in 1974. The share of the developing countries rose from 0.3% (0.2 million dollars) to 73.7% (401.2 million dollars) over the period. Again Mexico, Singapore and Korea have been the major beneficiaries, accounting for 25.6%, 20.0% and 16.7% of the total of the developing world respectively in 1974.[1]

Although textiles and garment were major items of import from developing countries under such tariff concessions in earlier years, electronics and machinery have become more important recently: they accounted for 56 per cent and 13 percent of the total U.S. imports from developing countries under Tariff Items 807.00 and 806.30 in 1972, respectively. In contrast, transport equipment, machinery and electronics amounted to 39 per cent, 27 per cent, and 18 per cent of such imports from developed countries, respectively.

---

[1] U.S. International Trade Commission, op. cit.

Sub-contracting from Japan

According to the Medium and Small-scale Enterprise Agency, the number of "bonded processing" contracts from Japan to developing countries increased by more than 40% from 1,315 to 1,847 cases, between 1969 and 1972. 52% of the manufacturer-parent firms and 85% of the trader-parent firms had local manufacturers as their sub-contractors in developing countries, the rest being subsidiaries of Japanese firms or joint-venture firms.[1] Again, geographical concentration is remarkable: in 1974, 126 (92.7%) out of a total of 136 Japanese-owned firms abroad which were processing imported raw materials for export, whether "bonded" or not "bonded" for duty-free import of such materials, were located in Asia, mainly Hong Kong, Korea, and Taiwan.[2]

As regards intra-firm transactions of multi-nationals, 21.1% of total sales of Japanese subsidiaries in Asia was directed to their principal firms in Japan. The percentage was much higher in some industries: 90% in furniture, and 48% in garment and wooden products. It was 27.5% in electrical and electronic machinery.[3]

RESULTS OF THE TOKYO SURVEY

With a view to supplementing the above-cited information, the chief expert undertook the already mentioned questionnaire survey in Tokyo. Questionnaire forms were sent to 416 members, and 157 firms (37.7%) replied, out of which 44 were medium and small-scale enterprises. They included 11 trading firms, one engineering consultancy firm and another firm which was out of operation at the time of the questionnaire.

Excluding these 13 firms, 58 out of 144 responding manufacturers (40%) had sub-contractors in developing countries (Table 2). 44 (30.6%) had been sub-contracting directly from Japan (cross-border sub-contracting) and 27 (18.8%) indirectly through their overseas subsidiaries (within-border sub-contracting). 14 firms had been doing both types of sub-contracting. 23 firms had been sub-contracting to or in more than one country, a striking case being Matsushita Electric Industrial Company whose subsidiaries were sub-contracting in 22 developing countries. Needless to say, these parent firms often had a large number of sub-contractors in individual countries. Matsushita, Philippines for example, had more than 100 local sub-contractors in December 1975, when the chief expert visited.

---

[1] The Medium and Small-Scale Enterprise Agency: *Chu-sho Kigyo Hakusho* (White Paper on Medium and Small-Scale Enterprises) for 1974, pp. 218–222.

[2] Ministry of International Trade and Industry: *Waga-Kuni Kigyo no Kaigai Jigyo Katsudo* (Overseas Business Activities of Japanese Firms), the Fourth Survey, 1975, p.79.

[3] Ibid., p. 178.

Table 2

International Sub-contracting to and in developing countries by Japanese Manufacturing Firms: Results of a questionnaire survey October 1975

(in no. of firms)

Industry	Total number of Respondents	Firms which had been doing international sub-contracting	Directly[1]	Indirectly[2]
Textile	18	5 (27.8)*	4	2
Electrical and electronic machinery	73	32 (43.8)	29	14
Transport machinery	14	9 (62.3)	2	7
Precision machinery	19	6 (31.6)	6	–
Other machinery	12	4 (33.3)	1	3
Other manufacturing	8	2 (25.0)	2	1
Total	144	58 (40.3)	44	27

* *The figures in parentheses indicate precentages in the total number of respondents.*

Note: 1. *Respondents who had been sub-contracting directly to firms in developing countries.*
2. *Respondents whose subsidiaries abroad had been sub-contracting to local firms.*

Source: *Tokyo Survey, October 1975.*

Geographical concentration in Taiwan and Korea was remarkable: 26 firms had been sub-contracting to or in Taiwan and 24 firms to or in Korea. They were followed by Thailand (12), Malaysia (8), Singapore (7), Brazil (6) and India and the Philippines (4 each). Out of the 103 references made regarding the country of sub-contractors, 88 were in Asia and 13 in Latin America. These figures are, however, not comprehensive even in relation to the limited number of samples, as there were firms which did not give a complete list of sub-contractor's countries.

In the list of countries of sub-contractors above, one notices a peculiar fact: not a single firm in the survey mentioned Hong Kong. This is incredible as it is common sense that Hong Kong is a country of sub-contractors. A plausible explanation seems to be that most of the "sub-contractors" in Hong Kong are foreign subsidiaries and joint-venture firms, and the respondents who are their principal firms did not regard them as their sub-contractors. It follows that the "sub-contractors" in the Tokyo Survey are "independent" from the parent firms, and do not include the respondents' subsidiaries and joint-venture firms.

The relatively small weight of textile manufacturers in Table 2 is also rather surprising. This seems due largely to the fact that the survey was conducted from Tokyo with a natural concentration of samples in this area, while the centre of the Japanese textile industry is in the Osaka area. Moreover, samples selected for the survey may be biased towards larger firms as the coverage of the Chambers' membership is obviously more complete among them than among smaller enterprises. It is known that a large number of medium and small-scale garment manufacturers, particularly in the Osaka area, have been sub-contracting to Hong Kong, Korea and Taiwan.

As regards the contents of sub-contracting, the cases of direct sub-contracting were almost equally divided between parts-manufacturing and production of end products. In contrast, indirect sub-contracting was almost totally related to parts-manufacturing, as expected.

Evidence from Sub-contractors' Countries

Manufactured exports from Hong Kong, Korea, and Singapore are largely involved in international sub-contracting.

In Hong Kong, where about 70% of the industrial output is exported, "most of the manufacturers produce according to orders received" from abroad, and it is "not uncommon" for the product to be distributed under the foreign buyers' brand.[1] A majority of overseas parent firms are traders

---

1 "Country Report: Hong Kong", p. 70.

which are engaged in commercial sub-contracting. Garment accounts for nearly one half of total domestic exports from Hong Kong. Cross-border industrial sub-contracting is observable in optical and photographic goods and electronics manufacturing,[1] but these two items accounted for only 3.5% and 10.2% of total domestic exports in 1975.

In Korea, too, export garment manufacturers, which earned $1,148 million, or 22.6% of total exports of the country, in 1975,[2] "do not produce until they receive orders from parent firms in foreign countries". On the average, each of them have about 10 parent firms.[3]

An interesting phenomenon, which is common to Hong Kong and Korea, and probably also Singapore, is that cross-border sub-contracting in electronics industry takes predominantly a form of intra-firm transaction between foreign firms and their subsidiaries or joint-venture firms. In Korea, for example, 74% of exports of this industry was attributed to foreign subsidiaries and joint-venture firms in 1975.[4] In Hong Kong, such firms' contribution amounted to 43.7% of total domestic exports of electronics products, as compared with a mere 3.4% in garments in 1974.[5]

Probably, this is due to the fact that electronics industry uses far more sophisticated technology and requires a greater care for quality control than garment manufacturing. It is also known that as the degree of sophistication of required technology increases, companies become more inclined to direct investment than licensing, to prevent leakage of know-how.[6]

As is clear from the above, with the exception of Mexico, cross-border international sub-contracting is largely concentrated in garment and electro-

---

1 Ibid., p. 72.

2 The Bank of Korea: *Economic Statistics Yearbook, 1976*, Seoul, 1976.

3 "Country Report: Korea" p. 107.

4 Ibid., p. 108. In this connection, it may be mentioned that, in a survey conducted by the Korea Development Institute in early 1974, nearly half of the 94 Korean partners who responded to the questionnaire had entered into joint-ventures with foreign firms for export purposes (cf. Chin Keun Park, "A Study of the Joint-investment in Korea," a paper presented at the KDI–HIID Conference on *Trade and the Blance of Payments in Korean Development*, Seoul, June 1974, p. 6.)

5 These percentages were calculated using domestic export figures in *Hong Kong Trade Statistics*, and data on establishments with overseas interest separately collected by the Commerce and Industry Department. The two sets of data may not be fully comparable, but should be good enough to provide a reasonably accurate sense of proportion.

6 Committee on Finance, United States Senate: *Implications of Multi-national Firms for World Trade and Investment and for U.S. Trade and Labour* (Washington D.C., Feb., 1973), p. 599.

nics industries in the Far Eastern developing countries and Singapore.[1] In India and Pakistan, this type of international sub-contracting is very much limited in quantity, although it may be considerable in terms of the percentage of total exports of individual industries.

Cross-border sub-contracting in transport equipment manufacturing has been a common practice among developed countries. Exports from developing countries are still limited mostly to replacement parts, particularly, for old model cars which are out of production in developed countries. As will be discussed later, there are all kinds of disadvantages which handicap automotive industries of developing countries as suppliers of assembly parts in international markets.

Nevertheless, there are a few encouraging signs on this front too. Daimler Benz of West Germany are now buying high tensility fasteners, engine moutings, oil seals, casting and forging from India. The total amount of orders is expected to be around 60,000 dollars in 1976.[2] Foreign subsidiaries in the Philippines have also been supplying assembly parts to sister factories in other countries, in order to meet regulations under the Progressive Car Manufacturing Programme.[3]

WITHIN-BORDER SUB-CONTRACTING

Republic of Korea

Sub-contracting has been growing important in Korean manufacturing. The percentage of sub-contractors' sales (i.e. sales of products manufactured on receipt of orders from other manufacturers) in the total amount of sales by the medium and small enterprises (enterprises with 5 to 199 workers) rose from 13.2% to 18.3% between 1968 and 1973.[4] The extent of this practice in different industries is indicated in Table 3 in terms of the percentage of sub-contractors in the total number of medium and small manufacturing enterprises, in reference to 1973.

All these figures, however, underestimate the importance of subcontracting in Korean manufacturing, because they cover only sub-contracting by manufacturing firms, and exclude sub-contracting by trading firms.

---

1 Western European countries' (notably West Germany's) imports under tariff concessions for off-shore assembly are small in amount and heavily concentrated in Eastern European countries and Yugoslavia (cf. Finger, op. cit., p. 367)

2 "Country Report: India", p. 88.

3 See Annex. I.

4 The Japan External Trade Organisation (JETRO): *Kankoku no Shitauke Kanren Kigyo* (Sub-contractors in Korea) (Tokyo, March 1975, mimeo), p. 4, quoting the Medium Industry Bank of Korea and the Ministry of Commerce and Industry, the Government of Korea.

Table 3

Medium and Small-Scale Sub-contractors in Korean Manufacturing (A partial appraisal, 1973)

Industry	(1) No. of small and medium enterprises[1]	Sub-contractors[2] (2) No.	Sub-contractors[2] (3) (2)/(1) x 100
Total manufacturing	22,881	5,088	22.2
Food and beverages	4,837	210	4.3
Textile, garment, leather	5,906	1,097	18.6
Wooden products, furniture	2,154	436	20.2
Paper and Printing	1,706	723	42.4
Chemical, petroleum, coal, rubber products	1,609	432	26.9
Non-metal mineral products	1,982	321	16.2
Crude metals	430	161	37.4
Metal products	1,383	454	32.8
Machinery	974	535	54.9
Electrical and electronic equipment	485	267	55.1
Transport equipment	775	329	42.5
Precision equipment	105	33	31.4
Other	635	90	14.2

Note: [1] *Enterprises with 5 to 199 workers only.*
[2] *A "sub-contractor" is defined to be an enterprise which manufactures its products upon receipt of orders from other manufacturing firms, but not from a trading firm.*

Source: *Columns 1 and 2 are based on data from JETRO, op. cit. pp. 15-18.*

This omission seems particularly serious in the light industries where commercial sub-contracting is prevalent. It is, for example, unrealistic that the proportion of sub-contractors in total number of enterprises was only 18.6% in textile, garment and leather industries: a similar proportion in Japan was as high as 79.5% in the textile industry and 69.6% in the garment industry in 1966. In the garment industry, trader-parent firms (i.e. wholesaler and retailer) were more common (46.5%) than manufacturer-parent firms (42.5%).[1]

73.6% of the sub-contractors depended on sub-contracting for over 60% of their sales, but only 11.5% had a single parent firm, 33% having two to five parent firms and the rest even more.[2] If the enterprises employing fewer than five workers had been included, the percentage of sub-contractors with a single parent firm must have been much higher, as is the case in Japan.

No separate statistics are available on within-border sub-contracting by foreign subsidiaries and joint-venture firms. Findings of the survey, however, suggest that at least some of them do have a fairly large number of local sub-contractors, although the amount of their sub-contracting is rather limited: the average number of sub-contractors was 73.4 in electronics (11 parent firms), 11.7 in machinery (5 parent firms), 86 in automobile (3 parent firms), and 103 in shipbuilding (1 parent firm), while the average percentage of sub-contracting in total material costs was 28.8 per cent, 18.8 per cent, 33.3 per cent, and 71.0 per cent, respectively.[3] As a similar percentage is between 70 and 80 per cent in these industries in Japan, the extent of within-border sub-contracting may be considered to be quite limited in Korea.[4]

Hong Kong

In Hong Kong, it is reported that 45% of manufacturing enterprises with over 20 workers were sub-contracting part of their work to other

---

1 Cf. Watanabe, op. cit., p. 55

2 JETRO, op. cit., p. 22

3 "Country Report: Korea", p. 110.

4 It is, however, necessary to note that the rate of sub-contracting in Japan is exceptionally high even among the industrialised countries. Although the six major automobile assemblers in Japan subcontract, on the average, over 70 per cent of material cost, the similar percentage is between 30 and 40 per cent in Europe and the United States. Only Volvo of Sweden depends on the external suppliers as much as the Japanese assemblers, but Volvo imports most of the parts and components from other European countries (cf. The Economic Research Centre, Kikai Shinko Kyokai: *Nichibei Kikai Kogyo ni Okeru Nai-Gai-Sei Mondai ni Kansuru Chosa Kenkyu* (A Study on Own-production and Sub-contracting in Japanese and American Machinery Manufacturing) Tokyo, May, 1975. pp. 128-130.

firms in 1973, paying, on the average, 3.2% of the gross value of their production to their sub-contractors. 45% of total sub-contracting firms belonged to textile and clothing industries, and 67% of the total sub-contracting payments was made by them. (Table 4).

This evidence seems to support strongly the view expressed in the Country Report that "within-border industrial sub-contracting is not common in Hong Kong."[1]

India

No systematic statistics are available on within-border sub-contracting by foreign subsidiaries in India, nor on sub-contracting among local enterprises. Scattered information, however, indicates that sub-contracting has been growing and that at least some of the large foreign subsidiaries have been sub-contracting to local firms fairly extensively and probably more than their counterparts in Hong Kong and Korea.

As regards the first point, attention may be drawn to the automotive industry. Here, local content of six major assemblers' vehicles rose from 32–70 per cent in 1960 to over 94 per cent in 1969,[2] and, although most of the local parts were produced by assemblers themselves in earlier year, they have become sub-contracted increasingly since 1960 and particularly after 1965, when the Government established a "reserved list" for parts-makers and prohibited the assemblers from manufacturing certain parts except with their existing facilities. In 1967, five studied assemblers bought out 40 to 62 per cent of their local contents.[3]

A recent OECD-sponsored study found that foreign subsidiaries in India use more local materials than those in other developing areas: nine foreign subsidiaries in India procured 60.9 per cent of their materials from local firms and only 15.6 per cent from their principal firms in 1970, as compared with 20.9 per cent and 67.3 per cent in the Far East (16 firms), and 43.9 per cent and 36.1 per cent in Latin American (25 firms)[4]. According to the Indian survey, Lucas TVS has some 300 sub-contractors on top of about 600 ordinary suppliers and Philips (India) has 561 sub-contractors. Dunlop, India seems to have been using a large number of sub-contractors for manufacturing sports-goods for export purpose. It is, however, difficult to tell how accurately one can form one over-all picture in this country on

---

1 "Country Report: Hong Kong", p. 73

2 Anne O. Krueger: *The Benefits and Costs of Import Substitution in India: a Micro-economic Study* (Minneapolis, University of Minnesota Press, 1975), p. 48.

3 Ibid., pp. 41, 55 and 56.

4 Grant L. Reuber *et al.; Private Foreign Investment in Development* (London, Oxford University Press, 1973), p. 153.

Table 4

Sub-contracting in Hong Kong (1973)
(Manufacturing Establishments with over 20 workers)

Industry	(1) No. of establishments	(2) No. of sub-contracting establishments	(3) (2)/(1) x 100	(4) Payment to Sub-contractors US$'000	% of gross output
Total	4,965	2,228	44.9	166,188	3.2
Textile	834	489	58.6	67,391	4.2
Clothing	1,250	510	40.8	44,450	3.9
Light metal products	500	284	56.8	13,060	6.6
Plastic products	762	328	43.0	10,895	2.8
Printing and publishing	204	145	71.1	6,755	5.4
Electrical and electronic products	375	129	34.4	6,582	1.1
Transport incl. shipbuilding	43	24	55.8	5,800	5.1
Other	997	319	32.0	11,255	1.4

Source: Industrial Production Census 1973, quoted in "Country report, Hong Kong," p. 71.

the basis of these pieces of evidence.[1]

Pakistan

The local expert in Pakistan reported that "organised sub-contracting has just started on both national and international levels" and that the extent of such practice is still "much limited."[2] Foreign subsidiaries import most of the materials, and the limited amount of their local procurement appears to take place mainly among themselves. In Sialkot City of Punjab, however, commercial sub-contracting has been in practice extensively in manufacturing of sports goods for the export purpose, involving 600 to 700 establishments, mostly cottage industries.[3]

Japan

The Tokyo Survey gained some information about relative frequency of within-border sub-contracting by Japanese firms (Table 2 and p. 8 are referred). Findings of a survey conducted by the Japan Export-Import Bank[4] throws some additional light on this subject.

The proportion of locally procured materials in the total supplies to the Japanese subsidiaries in developing countries (mostly in Asia) increases with the duration of their operation: while only 30% of the respondents with below two years' operation had managed to increase the local content ratio of their products since the commencement of their operation, 51% of the firms with eight years or longer operation had done so.

Such efforts were particularly remarkable in transport equipment, electrical and electronic equipment, and other machinery manufacturing, where 68% (23 firms), 64% (88 firms), and 49% (19 firms) had been increasing local contents respectively.

What is more interesting is their method of increasing the local contents. As Table 5 shows, such a result had been attained mostly by increasing purchase of existing local materials (41%) or own production (30%). At the same time, it is notable that 21 firms replied that they had successfully developed local sub-contractors by providing technical and/or financial assistance. As will be discussed later, parent firms send their own technical staff to sub-contractors, in order to give advice and show how to improve their efficiency and quality of their products. Sometimes they even send

---

1 "Country Report: India" pp. 94, 101.

2 "Country Report: Pakistan", pp. 119-120.

3 Ibid., p. 122.

4 Japan Export-Import Bank: *Waga-Kuni Kaigai Toshi no Genjo to Tenkai Hoko* (Current Situation of Japan's Overseas Investment and its Prospect in the Future), the Fourth Survey, (Tokyo 1974).

Table 5
Methods of Raising Local Contents Ratio by Jananese Subsidiaries in Developing Countries

Method	No. of Firms	% of Total
Total number of subsidiaries which had increased local contents since the beginning of their operation	345	100.0
1. Increase in purchase of existing local materials	142	41.1
2. Own production of previously imported materials	103	29.9
3. Purchase of previously imported materials from other Japanese subsidiaries	33	9.6
4. Improvement of quality of local firms' products by technical and/or financial assistance	21	6.1
5. Establishment of local sub-contractors by joint ventures	9	2.6
6. Establishment of local sub-contractors by Japanese sub-contractors	9	2.6
7. Other	28	8.1

Source: The Japan Export-Import Bank, op. cit., p. 147

sub-contractors, staff to their home country for training. As the standard of required technology goes up and as up-grading of sub-contractors becomes more difficult, own production and establishment of sub-contractors by joint ventures or by sub-contractors of principal firms seem to become more common, as has been the case in the Progressive Car Manufacturing Programme in the Philippines.

Chapter 3

Determinants of the Extent of International Sub-contracting

The extent and pattern of international sub-contracting in which the APO member countries are engaged vary from one country to another, as has already been discussed. Broadly, three sets of factors seem to determine them: (1) historical factors, (2) natural cost advantages resulting from resource endowments and geographical conditions, and (3) industrial and trade policies adopted by the governments of both exporting and importing countries.

In Pakistan where the history of industrialisation is relatively short, organised sub-contracting is still at its nascent stage on both national and international levels, as the local expert reported.

In India, where industrialisation started earlier than in almost any country in the region, a fair number of foreign subsidiaries have been in operation for decades, and some of them now have a large number of local sub-contractors. Moreover, the Government's import substitution policy obliges them to develop local procurement and sub-contracting. The existence of large local markets, however, keeps firms in India less enthusiastic about seeking cross-border sub-contracts. They are further discouraged in this respect by some industrial and trade policies of the Government which undermine the international competitiveness of Indian industries; e.g. by keeping the supply of raw and intermediate materials high in price and short in quantity, and the scale of operation far below optimum.[1]

In contrast, Hong Kong and Korea are more or less specialised in cross-border sub-contracting in labour-intensive light industries. This is obviously a combined outcome of such influences as the small size of domestic market, abundant supply of highly disciplined efficient labour, export-oriented industrial and trade policy, and advantageous location of the countries. There is no regulation that requires the use of locally-made parts and components.

[1] See, for example, Jagdish N. Bhagwati and Padma Desai: *India, Planning for Industrialisation – Industrialisation and Trade Policies since 1951* (London, New York and Bombay, Oxford University Press, 1970), Parts VI and VII; and Krueger, op. cit., p.56.

The situation was, and still is, much the same in Singapore. Now that the problem of unemployment has been solved for some time, however, the Government is trying hard to develop supporting industries to encourage within-border industrial sub-contracting.

Bearing all these points in mind, let us examine the motivation of international sub-contracting in more detail.

CROSS-BORDER SUB-CONTRACTING

The concentration of export-oriented labour-intensive light industries in Hong Kong, Korea, Singapore and Taiwan is no doubt attributable to the existence of abundant supply of efficient (i.e. cheap and well-trained) labour and the advantageous location of these countries. Their success, however, would have been much less spectacular without the Governments' encouragement, e.g. through schemes of export processing zones and bonded factories, and liberal trade policy. Bonded processing exports from Korea, for example, grew from 5.4 million dollars in 1964 to 1,102 million dollars in 1975, and their share in the total exports of the country rose from 4.5% to 21.7% over the same period.[1] The liberal trade policy adopted by Hong Kong and Singapore renders the entire territory of these countries a virtual processing zone.[2]

While these efforts of the governments of sub-contractors' countries greatly enhance their comparative cost advantages vis-a-vis developed countries, trade regulations such as Items 807.00 and 806.30 of the U.S. Tariff Schedules prompt would-be parent firms in developed countries to benefit from them.[3]

---

[1] *Economic Statistics Yearbook 1976,* op. cit. In 1974, 45% of bonded processing exports went to the United States, 27% to Japan and 10% to West Germany. Clothing and electrical, electronic and related products accounted for 34% and 33% respectively. (The Republic of Korea, Department of Customs Administration: *Statistical Yearbook of Foreign Trade*, December 1974.)

[2] Similarly, the phenomenal rise of Mexico as an international sub-contractor may be attributed safely to the Government's efforts aiming at the promotion of "border industries" and "in-bond industries" since 1966. For more information on these schemes, see "Fragmentos del informe de la Comicion de Aranceles de Estados Unidos sobre las industrias maquiladoras de exportacion", in *Comercio Exterior,* Apr. 1971, pp. 303-304, and Susumu Watanabe "Constraints on Labour-intensive Export Industries in Mexico", in *International Labour Review,* Jan, 1974, pp. 39-41.

[3] For a brief review of similar tariff concessions in other developed countries, see United States Tariff Commission: *Economic factors affecting the use of items 809.00 and 806.30 of the tariff schedules of the United States: Report to the President on investigation on No. 332–61 under Section 332 of the Tariff Act of 1930*, (TC publication 339), (Washington D.C. September 1970), pp. 9–32.

On the other hand, quota or ceilings imposed by developed countries on imports (notably, textiles) from Hong Kong, Korea and Taiwan have slowed down the rate of growth of international sub-contracting to these countries and tended to spread it over a wider geographical area. In the Bataan Export Processing Zone in the Philippines, for example, the chief expert saw a garment manufacturer from Hong Kong, who had invested there in order to overcome the U.S. quota on imports from Hong Kong. Another influencing factor was the introduction of a General Scheme of Trade Preference by the United States in January 1976, which would make exports from the Philippines more profitable.

Another important motive for cross-border international sub-contracting is related to the export incentives and obligations imposed by local governments. In Korea, imports of raw materials and items otherwise prohibited or restricted for sale in the domestic market is licensed, according to the export performance of individual firms. In India, foreign subsidiaries are required to export up to 60% of their output by governmental regulations. On the other hand, they can be exempted from the rule of maximum equity participation of 40% if their export performance has been good.[1] Some foreign automobile assemblers in the Philippines have been exporting parts (e.g. stamped car bodies in the case of Ford) to obtain licences for importing other parts and components which are essential for their continuing operation (cf. Annex I).

Findings of a survey conducted by the Indo-German Chamber of Commerce in 1973 (Table 6) illustrate the impact of such political pressure eloquently. In this survey, which covered over 100 Indo-German joint-venture firms in chemical, pharmaceutical and engineering industries, nearly 67% of the firms mentioned import entitlement, export obligations, or other similar reasons ((a), (b), (f) and (g) in the Table) as their motive of export.

Cross-border industrial sub-contracting can be a direct outcome of a plant-export contract, too. In 1971, a company in Bangalore was exporting electrical parts to a Japanese firm from which it had imported a plant. The deal had been so arranged as to overcome the problem of the too small size of local market and maintain the rate of operation of the imported plant at a reasonably high level.[2] Recently the Trade Development Authority of India has concluded a contract for transfer of a garment manufacturing plant from Germany, whereby the German firm will buy 75% of its output over a period of five years.[3]

---

1 "Country Report: India," p. 96.

2 Susumu Watanabe, "International Sub-contracting, Employment and Skill Promotion," in *International Labour Review*, May 1972, p. 432.

3 "Country Report: India," p. 86.

Table 6
Motives of Export
indo-German Joint-Venture Firms, 1973

		(In percentage of total number of surveyed firms)
(a)	To avail of import entitlement	39.0
(b)	To fulfil export obligations	24.0
(c)	Home demand is insufficient	13.4
(d)	To earn foreign exchange for the country	10.9
(e)	Exports are profitable	6.5
(f)	To get export incentives	2.2
(g)	To get sanction for renewal of the collaboration agreements	2.0
(h)	To enter into new Market	2.0

Source: *Indo-German Chamber of Commerce: Indo-German Joint Ventures, Export Situation (Bombay), a report of a survey conducted in April 1973, p. 12.*

WITH-BORDER SUB-CONTRACTING

Impacts of pressure from the local government appear to be even greater in the case of within-border sub-contracting. In the business world of industrialised countries, it is widely accepted that the cheapest way to supply their products to developing countries is to export them in the finished form and that the cost of production goes up and the quality deteriorates increasingly as the local contents ratio rises.[1] Import restrictions and regulations concerning minimum local contents, however, force foreign subsidiaries to sub-contract part of their work to local firms. The PCMP in the Philippines is a typical example (Annex I).

In India, the Government's distrust of the industrial "giants" and desire to restrict their further growth adds to the pressure to this end.[2]

Such motivation tells on the content of sub-contracting. Without

---

[1] For the best empirical study on this subject, see Jack Baranson: *Automotive Industries in Developing Countries,* World Bank Staff Occasional Papers No. 8 (International Bank for Reconstruction and Development, Washington, DC, 1969).

[2] "Country Report: India," p. 96. See also Nabagopal Das: *The Indian Economy Under Planning* (Calcutta, World Press Private Limited, 1972), p. 94.

a specific designation by the local authorities, the local content mix would be determined in such a way as to minimise the "cost – and quality-penalties", i.e. to keep the increase in the cost of production and the sacrifice of quality as small as possible. In the automobile industry, the assemblers try to meet this objective by choosing the following parts for sub-contracting:

(a) regular replacement items such as batteries, exhaust pipes, tyres, glass, etc. which have already been produced locally;
(b) items which require materials and production techniques similar to those used for existing local products of different use, e.g. seats, mats, linings, plastic parts, etc.; and
(c) items which are interchangeable among automobiles of different models.

SUB-CONTRACTORS' MOTIVES

It is almost tautological to say that any form of sub-contracting, either domestic or international, commercial or industrial, helps the sub-contractor to overcome his marketing deficiencies: the outlet for his product is guaranteed, while designs and specifications are supplied by the parent firm. Information about the location of would-be buyers and the changing pattern of consumers' taste is particularly important for exporters, and yet this is the hardest thing to get for the firms of developing countries. All the respondents in the Hong Kong survey and most of them in Korea acknowledged this merit of international sub-contracting.

Next to this, access to new technology and skills was most frequently mentioned in these two countries. The Pakistan Report, too, argues that the supply of technical know-how from the parent firm is a main advantage.[1]

However, assistance from the parent firms in regard to the supply of adequate and sufficient raw materials seems to be appreciated far more in India and Pakistan than in Hong Kong and Korea. No doubt this is due to the more serious shortage of materials and the restrictive import policy of the Governments in the former countries.

---

1 "Country Report: Pakistan," p. 123.

Chapter 4

Problems of International Sub-contracting

PARENT FIRMS' VIEWS

Difficulties which parent firms in the Tokyo Survey had been experiencing in sub-contracting to or in developing countries have been indicated in Table 7. The most common problem is related to the quality of products. This is followed by low technical standards of local firms, communication problems, unreliable delivery, and high costs or inefficiency. Poor quality of products, unreliable delivery, and high costs or inefficiency may be caused by poor technical standards (i.e. poor quality of technical staff and capital equipment) and/or inadequate supply of raw materials.

"Communication problems" arise from local firms' lack of experience in international business and consequent lack of understanding about buyers' needs, difference in business practices (managerial practices, labour administration practices, commercial practices, etc.) between the two countries, linguistic barriers, difference in work ethic, etc. The first cause of communication problems was also mentioned very frequently during the chief expert's visits to foreign subsidiaries in the survey countries, above all in connection with quality control and pricing of sub-contracted products. This point may deserve some elaboration.

To those manufacturers who have been working for the local replacement or repair parts market, for example, the assemblers' tolerance regarding the quality of their products is often too severe, and prices offered by them unacceptably low. This is not confined to the parent firms from abroad. An entrepreneur in an ancillary industrial estate in Bangalore once retorted to the chief expert's enquiry that why should he work for the [parent] company, while outside [in the local market] they did not complain of the quality of his products and were ready to pay more.

The situation is similar in the Philippines, where one of the major difficulties encountered by participants in the PCMP are related to the difference in pricing methods adopted by assemblers and repair parts manufacturers. Even in developed countries, repair parts cost much more than

Table 7
Problems Faced by the Japanese Parent Firms in International Sub-contracting to LDCs

(in No. of Firms)

	Textile	Electrical and Electronic Machinery	Transport Machinery	Precision Machinery	Other Machinery	Other Manufacturing	Total
Low technical standards	–	17	3	–	–	1	21
Quality control	3	10	4	2	1	4	24
Delivery	2	4	2	1	–	1	10
Financial problem	–	4	–	2	–	–	6
High costs or inefficiency	1	4	2	1	1	–	9
Inadequate supply of raw materials	–	6	–	–	–	–	6
Communication problems	2	8	–	1	–	1	12
Other	1	6	–	1	–	–	8
Total	9	59	11	8	2	7	96

Source: Tokyo Survey, October 1975.

assembly parts, due to the need of piece-by-piece packing, storage, etc. In the Philippines, the former cost, on the average, twice as much. Tolerance regarding the quality also differs. More than one entrepreneur high up in the hierarchy of the Automotive Manufacturers' Institute, who are supposedly encouraging sub-contracting, asserted that they would rather remain suppliers to the repair market than become regular suppliers to the participants in the PCMP, as the former was much more profitable. Obviously, if they were more growth-minded, their choice would have been different, but this seems not to be the case with most of them.

This picture contradicts the widely-held view that all units set up primarily for the replacement market are potential sub-contractors. Indeed, some of the interviewed managers from developed countries confided that, for the purpose of developing local sub-contractors, it was often better to establish joint-venture firms with local financial houses than with local manufacturers who were used to business in the replacement market. In the latter case, it was extremely difficult to make them understand the need for, and accept, new pricing methods and severe quality control.

Another point often made during interviews at foreign subsidiaries in this connection was the lack of quality-consciousness among local manufacturers, who tended to believe that for customers, "the cheaper the better".

All these findings are endorsed by the Indian Report, when it says that "most of the parent firms, particularly foreign subsidiaries, try very hard to find technically qualified entrepreneurs as their sub-contractors. In some cases, technically reliable employees are encouraged to leave the firm and become a sub-contractor",[1] and "one of the most common complaints about the local sub-contractors related to the lack of understanding of the needs of the customer, lack of initiative to improve or to modify and absence of quality consciousness".[2]

As already noted, the samples of the Tokyo Survey may be somewhat biased towards larger firms. Smaller parent firms may have different problems. Some light has been thrown on this question by the survey of the Medium and Small-scale Enterprise Agency (Table 8). Lack of skills, either for production or for designing, is still a primary problem. However, lack of information about local commercial practice is an even more serious problem to manufacturer-parent firms. The ambiguity and uncertainty of local governments' policy seem to be also threatening to smaller parent firms.

---

1 "Country Report: India," p. 91.

2 *Ibid.* p. 95.

Table 8
Problems of International Processing

(in % of total respondents)

Problem	Manufacturers	Trading firms
Ambiguity and uncertainty of local governments' policy	28	22
Lack of information about local commercial practice	48	23
Lack of skills	35	38
Inadequate capital equipment	–	6
Shortage of capital	14	25
Inadequate designing ability	7	22
Non-fulfilment of contracts	18	26
Linguistic barrier	21	9
Other	21	20

Source: The Medium and Small-Scale Enterprise Agency, op. cit., p. 226

Table 9
Means of Establishing Sub-contractors
Used by Japanese Parent Firms

(in % of total respondents)

The parent firms in manufacturing:	
Introduction by trading companies	39%
Overseas campaign (tour)	26
Introduction by foreign business acquaintances	17
Old trade partners	13
Other	5
The parent firms in commerce:	
Overseas campaign (tour)	37%
Old trade partner	23
Introduction by foreign business acquaintances	22
Introduction by Japanese trading companies	20

Source: The Medium and Small-Scale Enterprise Agency, op. cit., p. 222

Table 8 is also interesting in that it depicts clearly relatively strong points and weak points of manufacturers and traders as parent firms in international sub-contracting. As expected, the former are in a better position to overcome technical problems, but have much poorer information media than the latter.

Trading firms' abilities to collect information and their knowledge about products enable them to help other firms overcome trouble of finding suitable sub-contractors abroad: 39% of manufacturer-parent firms and 20% of trader-parent firms in the above-quoted survey acknowledge their help in this respect (Table 9).

Finally, in the Korean survey, ten out of twenty parent firms had problems pertaining to the supply of raw materials. This problem was followed by those related to poor quality of equipment and workmanship.[1]

SUB-CONTRACTORS' VIEWS

The problem of inadequate supplies of raw-materials is the most universal and it seems particularly serious where the import of raw-materials is tightly controlled. In India, the material cost of synthetic and blended fabric alone is much more than the garments of such materials can earn in export market.[2] In Pakistan, the situation appears even more desperate. Sub-contractors find it extremely difficult to comply with the orders, because raw materials of the desired specifications are not available in the country. Alternative raw materials are used and efforts are made to produce as good substitutes as possible, but they are very reluctantly approved even for trial purposes by the parent companies who are quality conscious.[3]

Similarly, restriction on the import of capital equipment can prevent local firms from receiving international sub-contracts, causing high production costs and poor quality of products. The garment industry in India, for example, does a large portion of cutting by hand, while other countries use automatic machines and even laser cutting machinery. As the India Report points out, this difference undoubtedly has an important bearing on the competitiveness of Indian products.[4]

In reference to cross-border sub-contracting, the India Report points out the problems of the Government's changing export policy and red tape of the customs office. They were also frequently mentioned during the chief expert's company visits. The seriousness of the first of the two problems has already been indicated in Table 8.

---

[1] "Country Report: Korea," p. 116.

[2] "Country Report: India," p. 95.

[3] "Country Report: Pakistan," p. 123.

[4] "Country Report: India," p. 95.

With her unique trade, industrial and international monetary policies, all these problems seem to be less important in Hong Kong. The survey found that the major problem is the quota imposed by industrialised countries on her exports. Rising prices of land, labour, and raw materials are another cause of difficulties here.[1] However, one may not be able to take this as a whole story, as the number of samples was so small (10 firms). Indeed, a survey on foreign companies which the Commerce and Industry Department conducted in March 1975 and in which 236 manufacturing establishments participated, 30 out of 58 establishments which reported they had operational problems mentioned difficulties in recruiting skilled workers. Twenty mentioned inadequate floor space or land, 6 mentioned difficulty in obtaining raw materials, and another 6 mentioned inadequate sub-contractors.[2]

In industries like automobile manufacturing, where economies of scale are considerable, the small size of local markets is a major constraint. It will prevent introduction of automatic machines which are required for severe quality control, or cause high cost of production by keeping the rate of operation of the machinery at an uneconomically low level. A director of the Korean Association of Automobile Industries frankly admitted that they just could not introduce some of the modern machinery which their overseas partners recommended, because the size of local market was too small. Similarly, an Indian delegation to the Japanese automobile industry reported: The limitations in capacity are such as not to permit the Indian Auto Parts Industry to entertain hopes of entering into arrangements with the vehicle manufacturers in Japan for supply of components as Original Equipment".[3]

In such an industry, the local government would have to choose one of the two alternatives: to wait for its development until a later stage of industrialisation of the economy, or to resort to extremely costly protective measures, e.g. tariff barrier or complete ban on the import, to protect the high cost local manufacturers. The latter case is fully illustrated by the PCMP in the Philippines. As Table 2 in Annex I indicates, the cost of a PCMP assembled car with the 1975 level of local contents cost, on the average, 750 dollars more than a completely-built-up imported car, net of all kinds of duties and taxes, margins and sales cost. Higher duties and taxes, however, kept the latter's retail price three times higher than the former. That this is a common problem in the developing world is testified by the

---

1 "Country Report: Hong Kong," p. 77.

2 Ibid., p. 78.

3 The Federation of Indian Chambers of Commerce and Industry: *Reports of the Delegation to Japan on Automobile Ancillaries* (March 1971) p. 10

already cited study of Jack Baranson.

All the country reports point out the difficulties local firms encounter in finding suitable parent firms. At the same time, it is noteworthy that 37 and 23 firms out of total 51 sub-contractors in the Korean survey complained about "delay in payment" and "unjustified cancellation of orders", respectively.[1] Why are these problems not mentioned in the other country reports? Is it because the other countries, specially India and Pakistan, have more urgent problems, or is it because sub-contractors in export processing are more vulnerable to these problems, or is it due to difference in the Government regulations? This is a point to be studied in the future.

---

[1] "Country Report: Korea," p. 116.

Chapter 5

Transfer of Technology and Skills from Parent Firms

IMPACT ON PRODUCTIVITY

International sub-contracting may have a favourable influence on the sub-contractor's productivity for different reasons. First, his productivity may go up simply as a result of fuller utilisation of the existing production capacity (capacity effect), which in turn may result from increased orders, or from better production planning that is made possible by a more regular flow of orders. The same effect may accrue from a better supply of raw materials, which is secured through the parent firm. Second, it may go up because of technical progress, i.e. introduction of new methods of production and management, and better training of workers (technology effect). An important component of this effect is the gain in efficiency which is brought about by the accumulation of experience, or by the "learning-by-doing" effect. Third, the sub-contractor's productivity may rise as a result of product innovation, i.e. introduction of new products or designs (product effect). It would, of course, be often impossible to attribute changes in productivity to only one of these effects, as they are, more often than not, inter-related.

In the Hong Kong Survey, 8 out of 10 sub-contractors claimed that their productivity had increased as a consequence of receiving sub-contracts, for one reason or another.[1] So did all the respondents in Korea. 52.8 per cent of total replies to the relevant question (144) were related to the capacity effect, 43.8 per cent to the technology effect, and only 3.4 per cent to the product effect, as Table 10 indicates.

It is natural to expect the relative importance of each effect to vary from industry to industry, and from country to country.

In the rest of the chapter, we will confine our discussion mainly to the technological effect.

---

1 "Country Report: Hong Kong," p. 75.

Table 10
Sub-contractors' Productivity Increase by Factor – Korea

(Number of replies)

Capacity effect:	
(a) increased sales	42
(b) regular flow of orders	32
(c) more favourable supply of raw materials	2
Technology effect:	
(d) improvement in quality of products	21
(e) better training of workers	16
(f) innovations in production processes	15
(g) decrease in production cost	7
(h) replacement of obsolete facilities	4
Product effect:	
(i) introduction of new products	5

Note: The total number of sub-contractors in the sample was 51.
Source: "Country Report: Korea", p. 115.

TRANSFER OF TECHNOLOGY AND SKILLS FROM JAPANESE FIRMS

Table 11 summarises findings of the Tokyo Survey on the technical assistance provided by Japanese parent firms. 39 out of 58 firms, which were directly or indirectly sub-contracting their work abroad, claimed that they had been providing technical assistance to their sub-contractors. Almost all the firms (36) had been extending assistance in regard to production techniques, 17 firms in regard to managerial techniques and 5 in the field of marketing.

Again, these findings may be supplemented with results of the Medium and Small-Scale Enterprise Agency's survey (Table 12), which also indicates that the most frequent assistance is related to production techniques, around 60% of parent firms being involved.

As expected, manufacturer-parent firms provided guidance regarding production techniques and production equipment more frequently than trader-parent firms. They also supplied more supervisory or advisory staff and financial resources. In contrast, supply of designs characterises assistance from trading firms. The difference between the two categories of parent firms in the Table may also reflect different patterns of technical and other assistance in textile manufacturing and various kinds of machinery manufacturing, as trading firms are dominant among parent firms in the former.

Table 11
Technical Assistance Provided by Japanese Parent Firms

(in no. of firms)

Industry	Production technique	Manage-ment	Market-ing	Other
Textiles (5)	5	2	1	–
Electrical and electronic (21)	18	11	3	2
Transport (5)	5	2	–	–
Precision (4)	4	1	1	–
Other (4)	4	1	–	–
Total (39)	36	17	5	2

Note: The figures in parentheses indicate the numbers of parent firms.
Source: Tokyo Survey, October 1975

Table 12
Assistance to Local Sub-contractors offered by Japan-based firms (1973)

(% of toal no. of firms)

Assistance	Manufacturers	Trading Firms
1. Continuous contract	50	55
2. Technical guidance	62	56
3. Managerial guidance	15	9
4. Supply of designs	27	48
5. Supply of staff	31	14
6. Provision of equipment	27	7
7. Financial assistance	12	4
8. Other	4	3

Source: The Medium and Small-scale Enterprise Agency op. cit., p.223
Note: Respondents sometimes gave more than one answer.

The minor weight of managerial guidance in these tables should not be considered as a sign of limited need of such assistance. On the contrary, most of the parent firms visited by the chief expert suggested that it was essential for the purpose of reducing the costs of sub-contracted parts and components and improving their reliability in quality and delivery. However, they added that, without capital participation, managerial guidance from parent firms were often impracticable, as local entrepreneurs regarded it as unnecessary intervention. Such resistance may be caused by a fear that managerial guidance from the parent firm might lead to a greater pressure for cost reduction by enabling it to control costing and revealing financial conditions of the enterprise, as is often argued in Japan.[1]

The most common pattern of technical assistance or guidance is the supply of technical advisers, regardless of the field of assistance. Almost any parent firm sends technical advisers on an *ad hoc* basis as such need arises: in the Tokyo Survey, 34 firms had been sending their technical personnel for the purpose of advising their sub-contractors (Table 13).

Table 13
Pattern of Technical Assistance Provided by Japanese Parent Firms

Industry	Licences and know-how	Technical advisers	Training at parent firms plant	Other
Textiles (5)	4	5	3	–
Electrical and electronic (19)	11	17	12	–
Transport (5)	2	4	3	1
Precision (4)	–	4	2	–
Other (4)	–	2	1	–
Total (37)	18	34	23	1

Note: The figures in the parentheses indicate the number of parent firms. Two firms did not specify the pattern of their technical assistance. This explains the gap in totals in this table and Table 11.
Source: Tokyo Survey, October 1975

[1] Cf. For example, Ichiro Oshikawa *et al.* (ed.) : *Kodo-Seicho Katei ni okeru Chucho Kigyo no Kozo Henka* (Structural Change in Medium and Small-scale Enterprises in a Process of Rapid Economic Growth) (Toyo Keizai Shimpo-sha, Tokyo, 1962), pp. 105–108.

TRAINING OF LOCAL PERSONNEL

Although the supply of technical advisers on an *ad hoc* basis is equally common among parent firms of any nationality, training of local personnel at the parent firm's plant appears to be more common among Japanese firms and European firms than among American firms. A tentative explanation is that American firms attract more trained people, e.g. by higher salaries and the company's name value. Where it is possible, they also tend to cover the lack of skills with machinery or technology,[1] as a government official in Singapore pointed out in an interview with the chief expert. At the same time, it is interesting to note that a number of American firms in the Philippines and India had recruited key personnel among graduates from American universities and trained them at their principal firms sometimes for several years on the job, before bringing them back home. Thanks to the large number of students who go to the United States, American subsidiaries seem to be in a better position in this regard than firms from other countries.

The figures on training in Table 13 are related to the training of indigenous sub-contractors' employees. As regards training of local workers employed by Japanese subidiaries, 36% of a total of 769 Japanese subsidiaries which responded to the question in the survey of the Japan Export-Import Bank in 1972–1973, had been sending their local employees to their principal firms' plants for training purposes.[2]

Many of these trainees go to Japan as part of the programme of the Association for Overseas Technical Scholarship (OTS), which, supported by the Ministry of International Trade and Industry, provides training curricula and facilities (e.g. dormitories), sets standards regarding trainees' treatment, and subsidises costs of training including air fares. 1,200 to 1,300 trainees benefit from this programme every year. About 75% are from Asian countries.[3]

Another 900 or so benefit every year from a special fund of the Japan International Co-operation Agency (JICA) and about 60 from the fund of the Japan Institute for the International Labour Organisation (ILO). Another 2,000 or 2,500 workers from developing countries are estimated to be trained in Japan every year without such support from the Japanese Government. It is, however, not clear how large is the share of training related to international sub-contracting.

---

1 Jack Baranson has recently discussed about the American practice of designing machinery in such a way as to minimise the problem of absorption capacity in developing countries (cf. the APO: *Intra-national Transfer of Technology* (Tokyo, 1976), p.5.

2 The Japan Export-Import Bank, op. cit., p.92.

3 *Annual Report* of the OTS for year 1975 (in Japanese).

From the company visits, however, the chief expert gained an impression that, among Japanese firms, the general trend is away from training of operational workers in Japan and that this is increasingly limited to foremen and supervisory staff and to pre-promotional stage. This is partly because, after all, training of a large number of workers in Japan costs a lot, particularly if the company must be prepared for their resignation shortly after the training course.[1] Some bitter experience in the past which was due to the cultural gap also seems responsible for this shift. (This point will be discussed below in reference to the report of a Singapore mission to Japan).

As regards the transfer of technology, as distinguished from skills, the major technology used by foreign subsidiaries comes, naturally, from their home countries. In the case of Japnaese firms, in 1974, out of a total of 1,041 firms in Asia, 69.5% were using key technology imported from their principal firms and another 25.2% from other Japanese firms.[2] The more sophisticated the technology becomes, the more eagerly the firms try to retain it within the firm by direct investment, rather than licensing. In the case of U.S. multi-nationals, transfer of technology to the rest of the world combined with capital investment accounts for over 75% of the total, in terms of payments and receipts on royalties and fees accounts.[3]

Information on such transfer of technology in the other types of international sub-contracting is very scanty. But the transfer of "small-scale production techniques" from Toyota to Delta Motors in the Philippines is an interesting example of transfer of "appropriate technology" (cf. Annex I).

EVIDENCE FROM DEVELOPING COUNTRIES

Cross-border Sub-contracting

In the Hong Kong Survey, four out of ten firms replied that they had not been receiving any assistance from parent firms, except designs and specifications. The remaining six had been receiving technical assistance regarding quality control including quality control staff in one case. Five firms had been supplied with production techniques and know-how. Thanks to such assistance, some of the responding sub-contractors had become confident enough to declare that they did not need any further assistance

---

1 Many companies make it part of their agreement with the trainee that the latter will work with the company at least a certain period of time (usually three years) after the training course. But this cannot be more than a moral constraint: apart from the question of the legal binding force of such a clause, the companies cannot expect much contribution from workers who stay with them half-heartedly.

2 The Ministry of International Trade and Industry: *The Fourth Survey*, op. cit., p. 162.

3 Committee on Finance, United States Senate, op. cit. p. 599.

from their parent firms.[1]

It is sometimes said that, cross-border international sub-contracting in assembly industries does not help technological progress of local industries. In reference to electronics, for example, a Korean author argues that the foreign firms bring semi-finished products for assembly in the labour-intensive segment of the production process and maintain a minimum relationship with local firms. Moreover, the local manufacturers of finished electronic products often import parts and components instead of using similar products made by foreign firms in Korea, technology inducement contracts with foreign firms often precluding the use of parts of other brands. Although the development of the industry has made a significant contribution to the overall growth of the economy and employment yet the current pattern is not likely to accomplish a full-scale diffusion of modern technology.[2] Korean country report more or less endorses such a view.[3]

This is, however, in a sharp contrast to the argument presented in relation to the same industry in Hong Kong,[4] which was initially engaged only in assembly of parts and components imported from the United States. While working in American subsidiaries, however, local engineers learned techniques of assembly line arrangements, batch and flow production systems, quality control, and supervisory methods, etc. Many of them started their own small firms, which now supply parts and components to big enterprises, replacing imports. For mutual co-operation and benefits, the latter assist them to overcome problems related to production techniques, supply of materials, and installation of machinery and equipment.

Even in Korea, one cannot neglect a number of impressive cases where cross-border commercial sub-contracting has encouraged technological progress not only in the immediatly related industry but also in other industries as a result of linkage effects. One striking example is *Shibori* industry in Korea.

---

1 "Country Report: Hong Kong," p.77.

2 Sang Chul Suh, "Development of a New Industry Through Exports: The Electronics Industry in Korea", in Wontack Hong and Anne O. Krueger (ed.): *Trade and development in Korea* (Seoul, Korea Development Institute, 1975), pp. 118–122. The productivity of this industry, however, improved remarkably between 1966 and 1972 (24.5% in labour productivity and 42.2% in capital productivity) (Chuk Kyo Kim, "Exports and Productivity Trend of the Korean Manufacturing Industries" idem., p. 134). This is largely due to the increased weight of the more productive foreign sector in the industry.

3 "Country Report: Korea," p.117.

4 See Fai-Tai Wong, paper in "Intra-national Transfer of Technology," Edited by APO, Tokyo, 1976, pp. 2–3.

Shibori is a silk material used for high-quality kimonos. Because of the mounting labour shortage, Japanese *shibori*-makers started sub-contracting the most labour-intensive process to Korea in 1962. At first only the tying of dots on imported silk fabric was done in Korea to be dyed and finished back in Japan. Gradually, however, cocoon-raising and spinning and weaving of silk material grew, and by 1975 most of the silk fabric used for *Shibori* processing was local. At the same time the dyeing industry gradually improved its technological standard and now does a considerable amount of the dyeing too.[1]

That *Shibori* is not a very exceptional case and this can be illustrated by the rapidly declining import content of Korean exports: imports for export processing accounted for 52 per cent of total exports in 1971, but only 29 per cent in 1975.[2]

All these pieces of evidence seem to suggest that distinction must be made between immediate effects and long-term effects when discussing technological impacts of international sub-contracting.

Within-border Sub-contracting

From scattered information one gains an impression that, in India, more efforts have been made by foreign subsidiaries for developing local sub-contractors and raising their technological standards. After decades of their operation in the country under the governmental regulations which require them to increase local content and sub-contracting to local firms, this seems to be the only natural outcome. A couple of most notable examples are quoted here from the Indian Report. Lack of systematic information, however, forbids attempt to establish an exact overall picture, and it is open to question how accurate this "impression" is.

Philips (India), which commenced operation in 1930, had 561 local sub-contractors in 1974. It tries out most of the locally sub-contracted jobs at its own factories beforehand. At the same time, three of their research and development laboratories assist the small scale suppliers in developing, testing, and evaluating prototypes. For the transfer of special know-how, experts are brought in from abroad. Philips (India) helps to finance tooling costs and train local suppliers' personnel. In manufacturing of such sophisticated components as single-sided printed circuit boards, which have very limited tolerance, Philips (India) supplied the technology to an Indian subcontractor, and trained supervisors and technicians for him. What is proba-

1 For a brief discussion on Korean Shibori industry, see Susumu Watabane, "International Sub-contracting, Employment and Skill Promotion", in *International Labour Review*, May 1972, pp. 433–434.

2 Cf. "A Study on Export Multiplier in Korea", in *Quarterly Economic Review* (Seoul, the Bank of Korea), Dec. 1976, p. 26.

bly more important, the Indian firm in turn passed the technology and experience on to its own sub-contractors, by training their staff, providing jigs, fixtures and dies.[1]

Another remarkable case is Ashok Leyland, a subsidiary which was established in 1956 by Leyland Motors, U.K. When the latter firm stopped manufacturing the "comet model", the former had to find a local source of supply of side members, which used to come from the U.K. Ashok Leyland obtained cold rolling technology from its principal firm and passed it on to a local metal pressing firm. It also helped the Indian sub-contractor in setting the roll correctly, choosing adequate materials, and controlling quality of the product. Japanese experts were invited to advise the sub-contractor, as raw materials were imported from Japan. It also arranged a visit of the sub-contractor's technical staff to a metal-processing company in the U.K. for in-plant training. In a similar way, Ashok Leyland transferred technology and skills to a number of other Indian sub-contractors for manufacturing tappets and valve guides from special cold cast iron, and crank-shafts.[2]

On the other hand, the Indian Report mentions about a foreign subsidiary which ceased its contract with a sub-contractor of ten years' standing in drawing and copper-coating of wires used in welding processes, for fear that the sub-contractor might become a competitor.[3]

As already noted, industrial sub-contracting is still in its early stage of development in Pakistan. As anywhere else, however, most of the parent firms provide all kinds of assistance to their sub-contractors, wherever sub-contracting is practised. To quote an example, Vauxhall Motors, U.K. provides National Motors with detailed drawings not only of components but also of tools needed for their production. The locally manufactured components are tested by Vauxhall, which also trains Pakistani engineers in the U.K. In return, National Motors offers all kinds of necessary help and technical know-how and sends advisers to its own sub-contractors. It also provides financial assistance, i.e. substantial advance for procurement of raw materials and development of dies and tools.[4]

The Pakistan Report also notes that international sub-contracting has recently started in printing industry in the country, after it introduced modern printing machinery following advice from foreign parent firms.

Before proceeding to next subject, it is necessary to note the spill-over effects of international sub-contracting on the technological stan-

---

1 "Country Report: India," pp. 101-102.

2 Ibid. p. 102-103.

3 Ibid. p. 94.

4 "Country Report: Pakistan," p. 126.

dard of local industries. This is largely due to the dispersal of foreign subsidiaries' ex-employees. The already discussed case of the electronics industry in Hong Kong illustrates the importance of such spill-over. The Indian Report notes that, sometimes, the inflow of ex-personnel of foreign subsidiaries has been a main factor contributing to the accumulation of technical staff in local industries, and that the movement of an experienced employee has been a primary cause of changes in technological standard of sub-contractors.[1] Similar effects are also recorded in the Pakistan Report[2].

PROBLEMS OF TECHNOLOGY TRANSFER

Of all the causes of difficulty parent firms experience in transferring technology and skills, the limitation in the sub-contractors' absorption capacity, which is largely due to inadequate basic education and training and lack of experience, seems to be the greatest. This is clear in the findings of the Tokyo Survey (Table 14). The India Report underlines this problem by saying: "Unless the sub-contractor has the basic ability to appreciate and absorb new skills and initiative to improve his abilities, the transfer process would become that of mere spoon feeding and remain inefficient".[3] That absorption capacity grows with experience is illustrated by a statement

Table 14
Problems Experienced by Japanese Parent Firms in Providing Technical Assistance

(in no. of firms)

Lack of basic education and training/absorption capacity	13
Inadequate return (e.g. royalty)	7
Difference in mentality	6
Linguistic barrier	6
High turnover of workers	6
Leakage of know-how and skills to competitors	2
Other	7

Source: Tokyo Survey, October 1975.

1 "Country Report: India," p. 94.

2 "Country Report: Pakistan," p.127.

3 "Country Report: India," p. 95.

made by a director of the Korean Association of Automobile Industries during the chief expert's visit: "Ten years ago we just did not know what to learn when we visited Japanese industries. Now we know better."

The importance of the question of "inadequate returns" and "leakage of know-how and skills to competitors"must be considerably underrated in the Table, as the respondents were those firms which had already been sub-contracting to local firms, instead of establishing their own plants: as already noted, these considerations are usually decisive in multi-nationals' choice between direct investment and technical tie-up.-

High turnover of trained workers is an inevitable problem in countries short of skills. With accumulation of experience in overseas business activities, Japanese firms are increasingly prepared to accept this inevitability and allow for loss of a certain percentage of trainees from the beginning.

The relevance of "difference in mentality" and lack of communication and mutual understanding to the technical assistance in quality control and managerial practices is clear from our earlier discussion. Lack of mutual trust is fatal to the transfer of managerial techniques and skills in particular. In the domain of training too, they can cause quite a serious problem which may develop into a problem of international politics. In this connection, the report of the mission of a Singaporean Industrial Training Observation Group to Japan[1] is full of lessons.

After consultation with the Economic Development Board of Singapore, a subsidiary of Nippon Miniature Bearing (NMB) had been training 206 girls and 44 boys at one of its principal firm's plants in Japan to make a quicker start of its plant in Singapore. A number of complaints had been filed by girls, however, regarding the conditions of life and work as well as the contents of the training scheme in Japan. This Mission was to investigate the matter not only at this particular plant but also in other companies.

Headed by the chief of the Manpower Development Division of the Economic Development Board of the Government of Singapore, the Group visited Japan between 23 October and 30 October 1973, at the invitation of the Japanese Chamber of Commerce and Industry in Singapore.

Regarding the conditions of work, it has been suggested that the company was making use of the Singapore trainees as "cheap labour" at 12,000 Yen per month, when the company had to pay 45,000 Yen to their Japanese workers. The Group found out, however, that the Singapore girls were given free food, free accommodation, free transportation,free medical care and accident insurance, while the Japanese were not given any remuner-

---

[1] *Report of the Industrial Training Observation Group from Singapore to Japan*, 8 December 1973, (mimeo.), made available to the chief expert by the head of the group after an interview.

ation in kind. Moreover, their air fare cost the company some 25,000 Yen a month per person.

Dissatisfaction with conditions of life came largely from difference in the pattern of life and customs in the two countries, i.e. regarding food and regulations of dormitories concerning friendship between young boys and girls.

As regards the contents of the training programme, the Group reports as follows:

> "Most of the trainees felt that after their eight months' training in Japan on return they could become supervisors or technicians straight away. Therefore, they were anxious to learn as much as they could even in areas outside what was planned for them. The company, on the other hand, felt that for the type of precision skills and quality work it needed, the Singapore trainees should concentrate all their efforts just on achieving maximum proficiency in only one process and become really skilled at it so that there would be zero defect in the product when these girls assume work on the production line in Singapore. The Singapore trainees obviously have not understood the concept of process specialisation in modern industry so the Group had to counsel them to accept the training concept the company had worked out. After they had mastered each process thoroughly diversification into other areas and skills could be more easily achieved. The Group noted that similarly in several cases of training in Germany, companies had similarly indicated to the Government that trainees had not understood the purpose and philosophy of industrial training for modern manufacturing and 'wanted to learn everything at the same time'. The Group concluded, therefore, that in the case of the NMB trainees this misunderstanding of training objectives, compounded by problems of adjustment of the workers to their *first* jobs and language and cultural difficulties developed some unhappiness manifesting itself in complaints on food, accommodation and wages."

The Group visited four other companies and held meetings at a number of institutes. One important point noticed by the Group was the fact that all trainees from abroad in training schemes supported by the Japanese Government go through a general orientation programme which provides them with basic information on Japanese work practices and attitudes, life styles and social customs as well as lessons of the Japanese language. Other training schemes lacked such an orientation programme, which was, in the Group's view, one of the major factors contributing to the unhappy experience of the NMB.

The crucial role of mutual trust and understanding in the process of transfer of technology is also illustrated by the case of transfer of "small-scale production techniques" from Toyota to Delta Motors in the Philippines. As noted in Annex I, this technology is a modified version of the engine production technique which Toyota used in the early 1950's to match the current scale of production. Considering the size of market in the Philippines, it was considered to be the most suitable production method, and yet, it is extremely difficult to persuade local industrialists to accept such old-fashioned production methods as they usually fancy up-to-date technology for the sake of prestige. "We were lucky that we had known Delta people for years and that they trusted us" so concluded the engineer who was responsible for this project and whom the chief expert interviewed at Toyota.

Chapter 6

Conclusion

SUMMARY OF FINDINGS

The bits of information collected through the survey indicate that various types of international sub-contracting have been practised extensively and growing fairly rapidly between firms in industrialised countries and those in some developing countries in the APO region. Endowment of resources (efficient labour in particular), geographical location, degree of industrial development, and size of local market are among the major determinants of the extent and the pattern of such business practice in individual countries. No less important factors are trade and industrial policies in the two countries concerned, but particularly those of the sub-contractor developing country.

Hong Kong, Korea and Singapore are characterised by the extensive practice of cross-border commercial and, to a smaller extent, industrial sub-contracting. These small countries are rich in efficient labour, and their trade and industrial policies are liberal and export-oriented, taking advantage of their geographical conditions which are convenient for international trade. In contrast, within-border industrial sub-contracting is dominant in India, where the Government's import substiution policy has been obliging major firms, and specially foreign subsidiaries to increase local content of their products and sub-contract part of their work to smaller local firms. The (potentially) large home market and the relatively long history of industrialisation – specially, the existence of a fair number of major foreign subsidiaries which have been in operation in this country for many decades – may favour such policy, but import substitution has often been achieved at the sacrifice of efficiency and therefore international competitiveness of local industries, thus discouraging cross-border sub-contracting. In Pakistan, sub-contracting is still new in practice and limited in its extent largely due to the short history of industrialisation, while in the Philippines it has started spreading, both within and beyond the national border, largely under the pressure of the Government's policy aiming at increasing savings and earnings of foreign exchange.

International sub-contracting can improve sub-contractors' productivity through various effects. In Hong Kong and Korea where cross-border sub-contracting is dominant, it has been enhanced, above all, as a result of fuller utilisation of existing production capacity. Technological progress is also substantial. It seems, however, natural to expect an even greater technological improvement in within-border industrial sub-contracting. There are a number of grounds on which this hypothesis can be supported. First, cross-border sub-contracting prevails in industries whose products are less precise (e.g. garments) and therefore require less efforts for technology transfer, while within-border sub-contracting in developing countries has been concentrated more or less in assembly industries where sub-contracted parts and components have to be made more precisely (e.g. auto-parts). Second, parent firms in cross-border sub-contracting go to a country where local firms and workers can do the work without too much assistance from them, while parent firms in within-border industrial sub-contracting have, usually, some local commitment in a form of invested capital and they are obliged to develop local sub-contractors when the local government introduces, or raises the standards of, import restrictions and local content requirements. Third, located in the sub-contractors' country, parent firms of the latter category are in a better position to provide technical guidance to local sub-contractors. Once sub-contractors of tolerable standards have appeared, however, the amount of parent firms' efforts will depend on the degree of competition they face in the local market as well as in the export market. To substantiate these hypotheses sufficient information could not be collected during the survey.

Apart from such comparative questions, most of the parent firms have been making considerable efforts to up-grade the technical standards of their sub-contractors, in any type of international subcontracting and in any country, as no developing country has reached an adequate standard yet. This is specially true in respect to quality control. Provision of tools, training of local staff, and supply of advisers for this purpose are common practice. Supply of designs and specifications, training of supervisors and foremen, supply of consultancy service regarding production process on the *ad hoc* basis are also taken for granted, although training of a large number of operational workers in the parent firm's home country appears to be on the wane.

The key factor in the transfer of technology and skills appears to be the sub-contractors' absorption capacity, i.e. the willingness and ability to learn. The size of this capacity in turn seems to be a function of the amount of their experience in business with industrialised countries. Only such experience can make them feel the need for learning new methods and prepare them for successful transfer of advanced technology and skills. In this sense, the element of time cannot be ignored, and the accumulation of experience in the cross-border sub-contracting may be a valuable process for development at a later stage. No less important is the degree of competition both the parties face in the local as well as international market. With a sufficient degree of competition, absorption of skills, techniques, and ideas for cost reduction and quality improvement will become a matter of life or death to the sub-contractors, while their parent firms' enthusiasm for technical guidance and assistance will also be increased.

PROSPECT OF FUTURE DEVELOPMENT

The rising trend of international sub-contracting between firms of developed and those in developing countries is likely to continue in the future. Indeed, no respondent in the Tokyo survey intended to reduce sub-contracting to or in developing countries, while 31 firms expected to increase it.

A notable discovery of the survey, as is evident from the Table 15, is that 8 out of 14 firms in automobile industry expected to increase sub-contracting. This is obviously related to within-border sub-contracting and

Table 15
Prospect of Development of International Sub-contracting in the Future

(in no. of firms)

Industry	No. of samples	No. of firms intending to increase international sub-contracting	No. of firms intending to maintain present level
Textiles	18	2	3
Electrical and electronic products	73	13	13
Transport machinery	14	8	1
Precision machinery	19	2	3
Other machinery	12	1	2
Other manufacturing	8	–	2
Total	144	31	24

Source: Tokyo Survey, October 1975

reflects the growing pressure from the governments in developing countries to increase local content.

In Korea, it seems only natural to expect a rapid expansion of within-border industrial sub-contracting, now that local sources of supply of basic materials, like steel, have been built up through the Third Five-Year Economic Development Plan. It is also obvious that, with the marked improvement in the labour market situation, Korean industries will shift towards more skill – and technology – intensive industries than garment manufacturing and assembly of imported electronic and electrical parts. The same is the case in Hong Kong, where under its Users Restrictions Scheme of Industrial Sites, the Government is encouraging capital-intensive and high-technology industries which produce, among others, raw materials for local industries. In Singapore, the current efforts made by the Government for the development of strong supporting industries (Annex II) are meant to boost within-border sub-contracting in that country.

Such changes in these countries are opportune, as cross-border sub-contracting in labour-intensive industries is more or less bound to spread over an increasingly wider geographical area, and continued rapid growth in the existing area (i.e. Hong Kong, Korea, Singapore and Taiwan) is unlikely. This is so, for a number of reasons. First, the spectacular success in these countries has been encouraging other countries to follow their example, by way of establishing export processing zones and introducing bonded factory schemes. The Philippines has already started partial operations of the Bataan Export Processing Zone. Both India and Pakistan are now contemplating "bonded processing" schemes, while the Electronic Export Processing Zone in Santa Cruz near Bombay is already in operation. A free trade zone for export garment manufacturing near Delhi is also under consideration. Related to these new developments is the second reason, i.e. quota imposed by industrialised countries on imports from the major exporting developing countries, the impact of which has been already discussed. The recent application of a General Scheme of Trade Preference by the United States has a similar effect. Last, but not least, rapid economic development in Hong Kong, Korea, and Taiwan has been improving the standard of living in these countries and the labour cost has been rising rapidly. This will also encourage a gradual shift of the centre of cross-border international sub-contracting in labour-intensive industries to elsewhere.

Expansion of international sub-contracting will, however, require increased efforts for transfer of technology and skills on the part of parent firms. Such need is clearly recognised by respondents in the Tokyo Survey (Table 16).

Table 16
Prospect of Technical Assistance by Japanese Firms in the near future

(in no. of firms)

Industry	Parent firms which intend to increase technical assistance	Parent firms which intend to maintain current level
Textiles (5)	2	3
Electrical and electronic machinery (31)	16	15
Transport machinery (8)	7	1
Precision machinery (6)	2	4
Other (4)	–	4
Total (54)	27	27

Note: The figures in the parentheses indicate the number of respondents.
Source: Tokyo Survey, October 1975.

POLICY IMPLICATIONS OF FINDINGS

(a) Improvement in the supply of raw materials

Shortage and poor quality of raw materials are among the greatest constraints on the developmnet of both national and international sub-contracting. Every effort should be made to solve this problem. Wider application of a preferential import policy, e.g. under the schemes of export processing zones and bonded factories, would deserve serious consideration, particularly in those countries where restrictive trade policies are in force. Establishment of a chain of raw material banks within a country also merits vehement support with a view to securing adequate supplies of such materials for small sub-contractors in particular.

(b) Improvement in quality control

The survey report is full of indications that quality control is the cause of the greatest headache on the part of parent firms, while strict quality control is considered as a major condition of success in sub-contracting and industrial development in the post-war Japan.[1] The first thing to be done in this front is to stimulate the quality consciousness among the local

[1] Cf. Jack Baranson: *Manufacturing Problems in India: the Cummins Diesel Experience* (New York, Syracuse University Press, 1967), p. 79, footnote 3.

industrialists. National Productivity Organisations and the APO should be able to play a key role in this task by organising seminars, lectures, and study tours, and by providing consultancy and other information services. Probably they can also run a national chain of depots of quality control tools and run testing laboratories, which are proposed in the Pakistan Report.

(c) Standardisation of products

Adoption of a national standardisation of parts and components will not only help quality control, but also promote division of labour and specialisation within an industry, which in turn simplify the worker' task, reduce the cost and duration of their training, and enhance their productivity by maximising the "learning-by-doing" effect through repetition of a narrow range of activities. A lively example is the Japanese sewing machine industry, which increased its production from a quarter of a million to 4.8 million units and its exports from zero to 3.2 million units between 1950 and 1969 by means of a thoroughgoing standardisation and specialisation in 1949 and 1950.[1] It is obvious that a high degree of standardisation of parts and components is much more difficult to attain in international sub-contracting, as different parent firms in different countries follow different specifications. Two alternative solutions exist: to have parent firms which follow a common specification, or to serve only one parent firm and try to get as much work from it as possible.

In certain industries, there is another alternative which is technically easier and economically necessary. This is a reduction in the number of models. Automobile manufacturing is a case in point. Everywhere in the developing world, this industry faces the problem of the small size of market, which prevents modernisation of capital equipment. And yet, it continues to produce a large number of models. In the Philippines, for example, 22 models are expected to be in production after three non-participants in the PCMP are "phased out", in the market of some 50,000 vehicles (including passenger cars, trucks, and "utility cars") a year (Annex I). If the number of models could be reduced substantially, efficiency in the industry would go up and quality of control would become easier.

(d) Pragmatic choice of technology

In modern industries, it is often technically impossible to produce

[1] Cf. S. Watanabe, "Sub-contracting, Industrialisation and Employment Creation", op. cit., p. 57, footnote 1. Before the adoption of the national specifications, most sewing machine manufacturers had been producing almost all their parts and components themselves, specifications varying from one enterprise to another. By 1963, the average number of parts produced by each manufacturer fell to 3 from 60 or 70 in earlier years.

internationally competitive products, in regard to price or quality or both, by old-fashioned techniques, as the Indian Report notes in reference to garment manufacturing. In order to overcome this problem while absorbing a maximum amount of surplus labour, it is necessary to differentiate the criteria of choice of technology, either by the market for which the products are intended or by the production process. The first approach has been adopted for example by the Korean embroidery industry where up-to-date machinery and old-fashioned second-hand machinery are used to manufacture exports and products for domestic market, respectively. The second approach has been a conventional strategy of Japanese industries, where capital investment is concentrated in the core processes of production, while a minimum amount of capital is used in the peripheral processes.[1]

(e) Financial support for improvement of facilities

In order to encourage such modernisation of capital equipment and level up the technological standard of sub-contractors and would-be sub-contractors to the international level, financial assistance from the local government may be necessary. Probably a substantial proportion of funds for medium and small-scale enterprises could be earmarked for this purpose. Such measures seem justifiable, particularly where a long contract is guaranteed by the parent firm after acquisition of the new machinery.

(f) Protection of sub-contractors

Implementation of such schemes as discussed under (e) presupposes observation of the contract by the parent firm. Legislation aiming at ensuring this and preventing abuse of sub-contractors might be considered, e.g. along the line of the Japanese Anti-Monopoly Law and the Payment of Sub-contractors (Prevention of Delay) Law. The Korean Government was contemplating introduction of such a law in 1971.[2] This kind of legislation may be considered at a national level, to start with.

(g) Information service regarding sub-contract market

In connection with the problem of finding suitable partners in international sub-contracting, the Pakistan Report mentions about the Sub-contracting Exchange, which the Pakistan Government is contemplating in collaboration with UNIDO. In India, the Trade Development Authority

---

1 See, for example, Gustav Rains, "Technology, Employment and Growth: Lessons from the Experience of Japan", in the International Labour Office: *Automation in developing countries* (Geneva, 1972). pp. 37–52.

2 JETRO, op. cit.

helps Indian firms to establish contact with potential parent firms abroad, just as the KOTRA in Korea. A weakness common to these institutes lies in the recruitment of officers who know individual firms and products well enough. In this regard, industrial associations, or sub-contractors' consortia proposed in the Pakistan Report, may be even more promising. These associations and consortia can, of course, undertake various other tasks to promote cooperation among sub-contractors in individual industries, including assistance in quality control, and procurement of raw materials. They may also be able to help strengthen sub-contractors' positions in bargaining vis-à-vis their parent firms or at a time of business recessions.

(h) Preferential incentives for transfer of technology

If core technology were to be transferred in international sub-contracting, it might be necessary for the local government to designate specific parts and components, rather than just indicate a local content ratio. If the list of chosen items turns out too ambitious for the existing standard of technology and skills in the country it will simply discourage international sub-contracting. In such cases, preferential treatment in regard to taxation, loans etc. might be granted to those parent firms which promise co-operation in developing sub-contractors in the related fields.

(i) Careful preparation of training courses abroad

As regards training of workers by parent firms, the following recommendations made by the already cited Singapore mission to Japan merit consideration:

(i) A comprehensive orientation programme should be organised by the responsible authorities of the country of the trainees together with the company for all trainees going abroad. Such a a programme should include language training, a briefing on work practices and systems in the receiving country, explanations of the training concept, career development prospects for trainees who are outstanding, etc.

(ii) As far as possible, future programmes of training abroad should be those for the development of higher skills or multiple skills to train personnel of supervisory, technician or engineer levels.

(iii) Should it be necessary to train workers abroad for first skills, however, special emphasis must be given to preparing such trainees both in their home countries and on arrival. The role for such preparation should be the joint responsibility of both the company and the local authorities of the trainees' home country.

(j) Increased efforts for cultural exchange

Findings of the survey indicate that "cultural gap" or lack of mutual understanding is one of the serious problems both in international sub-contracting and in transfer of technology and skills. Some companies which the chief expert visited had "cultural evenings" and language courses to overcome this problem. These might be better received, however, if they could be organised independently from individual companies. It might be recommended that National Productivity Organisations take initiative and organise such courses and gathering in collaboration with chambers of commerce and industry at home and abroad, and with associations of foreign subsidiaries. At the same time, governments might be recommended to expand their efforts for cultural exchange.

Annexure I

Progressive Car Manufacturing Programme (PCMP) in the Philippines

Car assembly in the Philippines started in 1951, when Fabar Inc. invested 2 million pesos to assemble Austin and Studebaker passenger cars. The number of assemblers rose in later years and reached the peak of 35 firms in 1964. Before January 1973, when the PCMP was implemented, there were no fewer than 19 companies assembling more than 60 different models of passenger cars.[1] In 1975, some 20,000 passenger cars, 20,000 trucks, and about 5,000 "Asian cars" were assembled by eight companies. When three non-participants in the Programme are "phased out", the remaining five assemblers will be producing 22 models.

The PCMP was announced by the Board of Investment (BOI) in May 1971 with a view to (1) saving foreign exchange through local production of automotive components, (2) creating work for existing small and medium enterprises and upgrading technological skill levels of local manufacturing industry. The end-product assemblers were to serve as the focal points through which engineering, management and financial assistance would be extended. It was also intended to (3) generate new export products as components of an automotive complementary programme in the ASEAN region.[2]

The second objective of the programme is related, as the BOI *Guideline* argued, to "the popular sentiment" against large enterprises. If a registered automotive assembler manufactured components within its plant, the other registered assemblers would not buy such components unless reciprocal relationships could be established. There were existing facilities which

---

[1] Automotive Manufacturers' Inst. Inc. (AMII hereafter): *The Philippine Automotive Industry*, Sept., 1974 (mimeo), p. 3.

[2] *BOI Progressive Car Manufacturing Programme* 1974, p. 2.

could manufacture many automotive components, if technical assistance and financial support were provided. On the other hand, it was expected that the facilities and technical skills employed by specialised component manufacturers could be effectively utilised for manufacturing components for a wider range of machinery and equipment than those of car assemblers. It was also argued that "in all highly industrialised nation, even with the tremendous volumes at which automobiles are produced, no automotive manufacturer is completely integrated, and many components are purchased from other manufacturers". Horizontal integration of industry is, therefore, preferred to vertical integration.[1]

In order to achieve these objectives, the Government established minimum domestic content requirements as follow[2] (Table 1):

Table 1

Target and Attained Local Content Ratios

	First Semester		Second Semester	
	Target	Attained	Target	Attained
1973	10%	15.9%	15%	23.6%
1974	20%	26.3%	25%	32.0%
1975	30%	34.5%	37.5%	–
1976	50%	–	62.5%	–

Targets for the subsequent years will be fixed after a review of the current programme. Only the assemblers registered with BOI under the programme were to get a foreign exchange allocation for the purpose of importing CKD units of cars and light commercial vehicles with engines of four or fewer cylinders and of 2,000 cc. or less cubic displacement.

These targets are to be achieved, while the following constraints are met: the foreign exchange cost for the local manufacturing of any component must not be greater than the cost of importing that component as part of the CKD unit; nor must the resulting peso cost be higher than the import cost including import duty, which amounts to around 50% ad valorem CIF. Seven companies applied for participation in the PCMP, out of which the following five companies were approved as registered participants:

1 BOI: *Guidelines for Progressive Car Manufacturing Programme* (Undated mimeographed document), Annex D.

2 Information supplied by BOI and AMII.

Company names	Foreign Licensor or Principal Company	Major Component	Car Model
Chrysler (Phils.)	Chrysler/ Mitsubishi	transmission	Colt
Delta Motor Co.	Toyota	engine block with cylinder head	Corona
DMC Inc.	Volkswagen	body and cast-machined vehicle components	Beatle Sakbayan
Ford (Phils.)	Ford	body	Escort
GM (Phils.)	GM, Isuzu	transmission	Holden

These five companies invested a total of 284.2 million pesos and employed 1,425 workers for this programme by December 1974,[1] attaining the average local content ratios indicated in Table 1. Value of their local purchases rose from 8.98 million pesos in pre-PCMP year 1972 to 32.10 million pesos in 1973 and 131.25 million pesos in 1974, and it is expected to reach 140.8 million pesos in 1975. The number of local component suppliers increasing from 80 in 1973 to 150 in 1974 and 220 in 1975, mostly as a result of diversification and enlargement of operation by firms which had been existing in the same or other fields. New firms accounted for no more than a few per cent.[2] Meantime, net savings and net earnings of foreign exchange amounted to 1.6 million dollars and 0.35 million dollars in 1973, 5.4 million dollars and 0.49 million dollars in 1974, and 3.7 million dollars and 0.8 million dollars in the first semester of 1975.

The increase in the local component content shown in Table 1 is, however, somewhat deceiving. The domestic content ratio here is given by

$\frac{A + B}{C} \times 100$, where

A = total value of components which are manufactured locally and used in the locally assembled car, calculated at the F.O.B. spare parts prices quoted in the official spare parts price list of the overseas supplier of the same model.

[1] *BOI Progressive Car Manufacturing Programme* 1974, p. 4.

[2] Information obtained during a conversation at the Parts Manufacturers' Association.

B = net foreign exchange earned by export of locally manufactured components (F.O.B. value of the exports less foreign exchange cost of manufacturing the components) attributed to the registered assembler during the period, and

C = the total F.O.B. export price of the same model completely built up but without extras and accessories, in the country of the overseas car supplier.

It is said that, on the average, spare parts prices are twice as high as assembly parts prices. Neglecting item B, therefore, one may be able to say that real domestic content ratio is about half the figures in the Table. As the value of B increases, real domestic content can decline even further.

In spite of the expressed preference of horizontal integration, the major thrust for increase in local contents has come from local production of new major components such as transmissions, bodies, cast-machined parts, and engine blocks and cylinders by assemblers, as well as from some degree of exchange of parts among assemblers (e.g. in the case of casting and forging). Purchase from non-assemblers is largely limited to such items as battery, tyres and other rubber parts, glass, textile and plastic parts including seats.

There seem to be serious constraints on the expansion of local purchase on both the buyers' and the sellers' side. On the buyers' side, the problem of paramount importance is "cost penalty", i.e. higher prices of locally-procured parts and components. Every assembler naturally tries to minimise the over-all cost penalty. Almost the only item that is cheaper than those of imports is seats. Even in this case, 30% import duty is the key element, and the question of quality is neglected. One of the assemblers who has not been exporting much (i.e. item B in the formula has a marginal value) reported a total cost penalty of some 900 pesos (120 to 130 dollars) to attain a local content ratio of 38%. This would mean that, adopting the local component mix chosen by this company, it cost about 24 pesos to increase local content by one per cent according to the Philippine formula, and nearly 50 pesos to increase *real* local content by one per cent. If similar calculations were undertaken excluding import duties, the cost penalty would become much larger.

On the other hand, parts makers, who have been supplying repair-parts market, are in a serious dilemma. Although they invariably insist that the assemblers should increase their local purchase, they are not enthusiastic about becoming regular suppliers for them. They prefer to remain in the repair-parts market, as they can make more money there and need not bother too much about quality of their products. In the repair-parts market, they enjoy a profit margin as high as 30% to 50%, while assembly-part prices sometimes do not cover their cost of production. It is difficult for them to meet assemblers' quality requirements with their handwork type of operation. The current scale of the market does not justify such a large capital investment which is needed to modernise their production methods.

The basic cause of difficulties on both sides, is without doubt, the small size of local market. It is important, in this connection, to note that, in developed countries, many of the major component makers started as part of assemblers' plants and became independent only after their technological base had become solid and their market large enough.[1] In this sense, the *Guideline's* statement that . . . "*even* with the tremendous volumes at which automobiles are produced, no automotive manufacturer is completely integrated" is rather misleading.

In order to raise technological standards of their workers and sub-contractors, each assembler has been making efforts. For training of their own staff, companies associated with Japanese firms, Mitsubishi and Toyota, had sent future key staff to Japan (four and twenty persons respectively) in the preparatory stage of their new transmission and engine plants. All the five PCMP participants invite engineers from their plants in Australia, Germany, or Japan to give advice to their workers and their sub-contractors, whenever need is felt. Their sub-contractors are also provided with financial assistance for tooling, support to get bank loans, and technical advice for management, quality control, on an ad *hoc basis.* The Japanese associates ask their "co-operative firms" (primary sub-contractors) to extend a technological tie to local sub-contractors for the purpose of improving their standards.

An interesting experiment attempted in connection with the PCMP is Delta's adoption of "small-scale production techniques" at its engine manufacturing plant. At a plant with a production capacity of 500 units a month, it was not appropriate to apply methods similar to those used in the current Toyota factories. A Toyota engineer in charge of designing the Delta plant realised that the latter's scale of production was approximately the same as that of Toyota before 1950. Accordingly, the design was prepared modifying Toyota's production facilities in those days. At the time of test operation of the plant facilities in Japan, twenty engineers and mechanics were brought from the Philippines to learn how to operate them.

Apart from individual efforts, these assemblers organise common training programmes for their sub-contractors, jointly with BOI and AMII. For example, in a month-long training course (seminar) in September–October 1975, which covered quality control, production planning and control, and methods improvement and waste reduction, 41 people from

---

[1] On this point, see Toyota: Automobile Manufacturing Co.: *30 Years' History of Toyota Automobile Manufacturing Co.* (1967) (in Japanese).

36 local component manufacturing plants participated.

A notable feature of this PCMP is the non-existence of any material support from the Government. There is no subsidy nor tax concession granted to assemblers and parts manufacturers, the only incentive that exists is the allocation of foreign exchange for import of CKD units. Another kind of adjustment in favour of assemblers is related to the tariff barrier: the import duty on CKD import was raised from 25% in pre-PCMP 1972 to 30% in 1975, and that on CBU import was 100% of CIF price (Table 2).

Table 2
Comparative Data on Car Prices

	Assembled Pre-PCMP Year 1972	PCMP Assembled 1975	Imported CBU 1975
FOB	$1,324.00	$1,556.50	$1,677.00
Freight	110.00	105.00	250.00
Insurance	2.00	2.50	3.00
CKD price, CIF	$1,436.00	$1,664.00	$1,930.00
	₱9,335.00*	₱12,480.00*	₱14,475.00
Customs duties	2,334.00	3,744.00	14,475.00
Advanced sales tax	1,021.00	1,419.00	28,950.00
Cost of local components			
Tyres and batteries	800.00	1,440.00	–
Other components	–	5,060.00	–
Assembly labour and overheads	750.00	1,200.00	–
Administrative and selling expenses	950.00	1,500.00	1,200.00
Final sales tax	2,260.00	2,416.00	59,350.00
Assemblers margin	2,803.00	2,500.00	2,500.00
Dealers margin	5,717.00	6,971.00	6,700.00
Suggested retail price	₱25,970.00	₱38,730.00	₱127,650.00

**1972 figure includes cost of completely knocked down (CKD) car minus tyres and batteries. 1975 figure includes cost of CKD car minus tyres, batteries and local components produced due to PCMP.*

Source: Press release by the AMII, dated 18 September 1975

In view of the still very low *real* local content ratio, it is probably difficult to justify that, as critics of the Programme assert in the Philippines, the Programme has been a failure because they have led to tremendous price increases: price increases are attributable mainly to increased duties and taxes, dealers' margins, import prices (a considerable part of which is due to changes in the foreign exchange rate (cf. the gap between price indices of CKD in terms of US dollars and Pesos in Table 2).

It is too early to assess the degree of success (or failure) of the Programme in terms of transfer of technology and skills. As sub-contracting to local firms is still confined mainly to textile and rubber items, and as sub-contracting of skill-intensive work (e.g. casting and forging) is limited largely among assemblers, however, it does not appear that a lot has been achieved regarding objective (2) of the Programme. Nor is progress impressive as regards the objective of developing a horizontal, rather than vertical, integration. The fundamental cause of difficulty appears, as already mentioned, to be the extremely small size of local markets. One might well question whether efficient component manufacturing industry could be developed economically, as the *Guideline* envisages, *before* securing a certain size of market e.g. by a regional complementation agreement.

Another point which is worth studying is the fact that assemblers sometimes prefer to set up a new parts manufacturing factory jointly with local capitalists, rather than existing repair parts manufacturers, due to the latter's lack of cost- and quality-consciousness. In this connection, it is noteworthy that industrial progress in developed countries has often been achieved as a result of replacement of old-fashioned enterprises by new progress-minded ones. Validity of the idea of continuous evolution of efficient parts-makers out of existing repair-parts makers, which is implicit in the *Guideline*, deserves a serious re-examination.

Annexure II

Encouragement of Transfer of Technology and Skills for Supporting Industries in Singapore

Since its separation from Malaysia in 1965, Singapore's economy has maintained a high rate of expansion. Its GDP grew 11.9% per annum up to 1974, and gross value added of the manufacturing sector alone about 17% a year during the period (both in constant prices). Needless to say, such a rapid economic expansion has been made possible by a vast amount of export-oriented foreign investment. Recently the Government has been trying to attract foreign firms in technology- and skill-intensive industries. In doing so, however, the local availability of supporting industries, e.g. in casting, has been questioned. Another problem is the supply of skilled workers. To meet these requirements, the Government has introduced a number of schemes. Their histories are much too short for an evaluation of their achievements, but these schemes are of some interest to us, as they are more or less directly intended to promote domestic and international sub-contracting and improve skill levels of involved workers.

CAPITAL ASSISTANCE SCHEME

Towards the end of 1974, the Government established this scheme to provide equity capital and loans at attractive terms to assist the establishment of such small to medium companies industries, that "will introduce new skills and manufacturing technology, upgrade and diversify existing operations, and bring about valuable integration in manufacturing processes".[1] Up to 50% of the required equity capital can be taken up by

[1] Economic Development Board: *Capital Assistance Scheme* July 1975.

the Economic Development Board. Since the EDB's intention is not to participate in industry but to facilitate its establishment by lessening the risk burden on the investor, the company can buy back part or all of the equity which the EDB holds, as a general principle, any time during the first five years.

As the attached list of the "priority development" products in Annexure III suggests, this scheme can be regarded as an important thrust aimed at promoting local production of hitherto imported parts and components and that of sub-contracting, both domestic and international.

OVERSEAS TRAINING SCHEME

Under this scheme, the Government (EDB) offers financial assistance for such training programmes of companies that are intended to introduce new skills and techniques or to supplement local training capacities during peak periods. EDB officers, stationed in the United States, Europe, and Japan for the purpose of promoting skill-intensive and high-technology investment, will place selected Singaporeans in companies which are likely to invest in Singapore. They are enrolled, preferably, in programmes leading to such standards of apprenticeship or other skill levels which are nationally accepted in respective countries. The idea is that when firms come to invest in Singapore, they will be assured of a supply of skilled workers and technicians trained to their national standards or to their own needs.

Financial assistance may also be given to an existing company which is launching an approved overseas training programme to introduce new skills and techniques. New companies may also use the scheme as a "quick start" programme to prepare key personnel. The basic idea is that in accepting Singapore trainees the participating company would bear no more expense than if these were their own trainees. Any additional costs such as return air fares, orientation expenses, warm clothing allowance, and a cost of living subsidy, will be paid by the EDB.

Over 200 trainees were sent abroad during the three years preceding the end of 1975, when 185 trainees were abroad under this Scheme.[1]

Trainees in Germany follow 3 to 3.5 years courses leading to the *Facharbeiter* certificates for highly skilled workers. The main fields of training are machine building, precision mechanics, tool and die making, electromechanics and pattern making. Those who go to Switzerland are trained, above all, in automatic lathe operation, tool and die making, and precision machining, for 1 to 1.5 years. The programme in Japan covers a wider field of technology and includes the manufacture of watches, indicating instruments, springs, oil seals, as well as machine mould making and

1 Singapore Economic Development Board: *Annual Report 1974–75*, p. 29.

foundry operations. Pilot programmes are also starting in Denmark, Canada, the U.K. and Australia.

All trainees are "bonded" to the EDB or directly to the company (subject to the approval of the EDB). The participating companies are given the first option to employ all or any of the trainees in Singapore. If they are not operating in Singapore, the trainees will be released to other companies (excluding direct competitors if so pre-arranged). Should they subsequently have a need to employ similar skilled personnel in Singapore, maximum assistance will be given. Non-participating companies which are expanding their industrial activities in Singapore may also apply to the EDB to utilise this source of skilled workers. In such a case, a release fee will be charged.[1]

INDUSTRIAL TRAINING GRANT SCHEME

Local and foreign companies operating in Singapore which organise acceptable training programmes to upgrade the level of unskilled or semi-skilled workers into skilled workers or craftsmen, may negotiate with the EDB for training grants. The grants are awarded to encourage companies to expand or start 2 to 4 year training programmes aimed at meeting their own manpower requirement. Companies intending to integrate or upgrade their operations are also encouraged to conduct short upgrading courses. The scheme allows for grants ranging from 3,000 to 9,000 Singapore dollars[2] for each worker. Qualifying skills range from precision toolmaking to pattern making. In 1974–75, over 6.2 million Singapore dollars was committed for the training of 1,205 workers at 15 companies.[3]

An interesting variation is the training programme of Mitsubishi Singapore Heavy Industries (PTE.) Ltd. The company was established in April 1973 and its dock was planned to start its operation by August 1976. What is unique about their approach is the adaptation of the company's training programme to the existing public vocational training programme of the Government.

The first group of 41 workers were recruited among those who had just completed courses at formal vocational training institutes. They followed a six-month (1,012 hours) off-the-job training programme designed by the MSHI according to its own need, and then a 2-year on-the-job training course. The subsequent groups, recruited among school-leavers, follow one year's standard programme of the Institute and another year's MSHI programme.

[1] Singapore Economic Development Boarcd: *Information Sheet No. 74.006,* dated 23 July 1974.

[2] US$1 = 2.3 Singapore dollars.

[3] EDB: *Annual Report 1974–75*, pp. 31–32.

In both programmes, existing training facilities and instructors are deployed. The MSHI, however, sends Japanese engineers (16 of them between January and October 1975) for the purpose of teaching its trainees and also levelling up the standards of the Institute's instructors. It also contributes machinery and tools as training materials. Vocational training institutes in Singapore are said to be equipped excellently, reflecting the Government's enthusiasm. However, local training staff is not adequate, both in quality and in quantity. The MSHI's programme is considered by the Government to have a great merit in that it contributes not only to local skill supply but also to the improvement of local teaching staff.

JOINT TRAINING CENTRE

Under this scheme, the centres transfer technical know-how, training soft-ware and instructional methods to a corps of local staff most of whom have previously been trained abroad, over a five to seven year period. The training centres conduct the first two years, and then the apprentices are placed in industry for on-the-job training. The Tata-Government Training Centre and the Rollei-Government Training Centre started their operation in August 1972 and in June 1973 respectively. Enrolment stood at 167 and 215 each as on 31 March 1975. The Philips-Government Training Centre started in July 1975 with 80 trainees.

Each company provides teaching staff, while the Government (EDB) supplies training materials and facilities, as well as trainees' wages while they are at the training centre. The companies can retain those trainees whom they want and the latter are obliged to work with them at least three years after the completion of their training. The Government is committed to provide all the trainees with a job. No provision seems to exist for potential left-overs. In the case of the Tata-Government Training Centre, EDB is taking over the centre after October 1977.

Annexure III

List of Priority Development Products in Singapore

1. **Metal Engineering and Machinery**

   Engines and turbines – internal combustion engines, including diesel, for various automotive and static applications; steam, gas and hydraulic turbines; generator sets.

   Metalworking machinery – cutting and forming machine tools, tools and dies, die sets, jigs and fixtures; precision measuring devices and machine tool accessories; machine parts.

   Special industrial machinery for food processing, textile industry, commercial laundry; woodworking, timber processing and extraction; printing, bookbinding, typesetting, platemaking; automatic vending machines, measuring and dispensing pumps.

   General industrial machinery – pumps, compressors, fire sprinklers; ball and roller bearings, power transmission systems; blowers, exhaust, pollution control equipment; furnaces and ovens; portable power tools and hand tools.

   Office machinery – typewriters, cash registers, addressing machines; duplicators and copiers.

   Construction, materials handling and other machinery – bulldozers, tractors and other construction machinery; elevators, escalators, conveyors; hoists, cranes; forklift trucks, stackers, industrial tractors and trucks; mining and oil field machinery, farm machinery and equipment.

2. **Aerospace Engineering**

   Aircraft, hovercraft and gliders.

   Aero-engines, propellers and parts.

   Aircraft parts and auxiliary equipment, instruments, electrical, electronic and hydraulic equipment.

3. **Shipbuilding and Transport Equipment**

Bulk carriers, gas and oil tankers.
Container ships, combo, roll-on/roll-off and LASH vessels.
Shipping and aircargo containers.

4. **Ship Machinery and Fittings**

Marine engines, boilers, heat exchangers, out-board motors, drive shaft, propellers, winches, davits, windlass, hatch covers, doors, window, marine hardware, life-boats and other safety equipment machine parts (cylinder liners, pistons and piston rings, white metal bearings) hydraulic systems.

5. **Optical Instruments and Lenses**

Optical glass, lenses, prism and lens systems.
Microscopes, telescopes and binoculars.
Optical testing and measuring instruments.
Medical and opthalmic instruments.
Photographic equipment, parts, attachments and accessories, still and motion picture cameras and projectors.
Photocopy and microfilm equipment.
Photographic and X-ray films.

6. **Engineering, Scientific and Medical Equipment**

Professional, scientific and laboratory instruments for nautical, navigational, aeronautical, surveying, drafting, laboratory work and research.

Measuring and controlling instruments for temperature, pressures, fluid flow, physical and chemical properties; dial pressure gauges; chemical balances and scales.

Automatic process controls and systems.

Surgical and medical equipments; surgical instruments; artificial kidney machines, heart-lung machines; incubators; orthopaedic, prosthetic and surgical appliances.

7. **Electrical and Electronic Products**

Electric motors, generators, transformers.

Switch-gear and switch-board and control equipment; automotive equipment; electric household appliances, portable tools, shavers. Electrical and electronic measuring instruments; electronic process and machinery control equipment and systems.

X-ray apparatus; medical electronic equipment for diagnosis, intensive care and therapy.

Hearing aids, heart pacemakers.

Electronic flash guns.

Electronic data processing equipment– computers, peripheral, terminal and inter-face equipment; electronic calculators and accounting machines.

Communication, signalling and detection equipment, including broadcasting equipment, radar; navigational and guidance systems.

8. Basic Chemicals, Petrochemicals and Fine Chemicals

Methanol, 2-ethyl hexonol and highee alcohols.
Ammonia, urea.
Carbon black.
Benzene, toluene, xylenes.
PTA/DMT, caprolactum.
Polyester, nylon, acrylics and other synthetic fibres.

Polyenthylene (high and low density), polypropylene, vinyl chloride monomer/polyvinyl chloride, styrene monomer/polystyrene, ABS, vinyl acetate monomer/polyvinyl acetate, and other thermoplastics, alkyds, polyurethanes, unsaturated polyester resins, phenolic, melamin, urea, epoxy and other thermosetting plastics.

Silicones.

Dyestuffs, catalysts, pharmaceuticals and other fine chemicals.

Phthalic anhydride, DOP and other plasticizers.

PART II: COUNTRY REPORTS

Chapter 7

Sub-contracting in Hong Kong

Hong Kong Productivity Centre

SECTION I: GENERAL REVIEW

This report consists of two sections. The first section is a general review of international sub-contracting in and to Hong Kong, which was prepared on the basis of readily available information. The second section is an analysis of findings of ten case studies on selected manufacturing companies: three in precision equipment (watches, cameras and optical equipment), four in electrical and electronic appliances and machinery, and three in garments. This analysis is supplemented with findings during interviews at governmental offices and other commercial and industrial organisations.

The survey was conducted from January through May 1976. The ten case studies involved in-depth personal interviews with the management of the selected companies. On the average, a case study took two to three visits to collect information on, e.g., production, finance, and marketing departments. Five of the sample firms were joint-ventures with foreign companies, and four were totally owned by foreign companies, the remaining one being a pure local firm. It is noted that the findings of the report are based on the ten case studies and any generalization may be limited in scope.

EXTENT OF THE CURRENT PRACTICE

Manufacturing Industries and Foreign Investment

In 1973, the Census and Statistics Department carried out an Industrial Production Census of Hong Kong. Total output of manufacturing establishments with over 20 workers in that year was US$5,165 million. Three industries (textiles, clothing, and electrical and electronic equipment and related products) accounted for 65 per cent of the total (Table 1).

Table 1
Industrial Production in Hong Kong (1973)[1]

Industry	No. of Establish-ments	Gross Value Added (US$ mil.)[2]	Per Cent of Total
Textiles	834	1,612	31.2
Clothing	1,250	1,138	22.0
Electronic and Electrical Products	375	598	11.6
Light Metal Products	500	385	7.5
Plastic Products	762	385	7.5
Food Processing	132	209	4.0
Printing and Publishing	204	126	2.4
Transport equipment (including shipbuilding)	43	113	2.2
Photographic and Optical Goods (including watches and clocks)	82	73	1.4
Chemicals	74	67	1.3
Paper and Paper Products	96	62	1.2
Non-electrical Machinery	77	41	0.8
Wood and Cork	56	37	0.7
Rubber Products	60	30	0.6
Furniture	56	26	0.5
Footwear	46	21	0.4
Leather	37	18	0.3
Others	281	224	4.4
Total	4,965	5,165	100.0

Note: 1 *Establishments with over 20 workers only.*
2 *H.K.$5.085 = US$1.00*

Source: *Census and statistics Department, Government of Hong Kong, Industrial Production Census, 1973.*

Hong Kong's manufacturing industries are heavily export oriented. In 1973, domestic exports amounted to US$3,831 million, or 65 per cent of the total production. They grew to US$4,546 million in 1974 and US$4,627 million in 1975. The above-mentioned three industries accounted for 65 per cent of the total exports in 1975.

According to the Commerce and Industry Department (C & I.Dept), there were 271 industrial establishments with overseas interests in December 1975. They employed 59,607 workers, or 8.8 per cent of the total workforce in the manufacturing sector. Total value of foreign investment in these establishments was estimated to be US$343 million.As shown in Table 2, investment in the already cited three industries and precision equipment industry amounted to two thirds of the total.

In terms of the amount of capital investment and the number of establishments, the United States, Japan and the United Kingdom were the most important investors. In December 1975, there were 99 establishments with U.S. interest, with capital investment of over US$162 million (or 47.2 per cent of the total). Japan, with interest in 72 establishments, was the second largest source of overseas investment, with capital investment of US$53 million (15.4 per cent). Capital investment by British firms amounted to US$32 million (9.4 per cent).

The United States is the leading overseas investor in electronic and electrical product industry, and chemical product industry. Japan is a major overseas investor in textile industry, printing and publishing and allied product industry as well as in electrical product industry.

In March 1975, the C. & I. Dept. carried out a survey on foreign interests in Hong Kong. A total of 236 manufacturing establishments participated in the survey. According to its findings, 118 out of 236 (or 50 per cent) were totally owned by overseas interests (including 78 subsidiaries of overseas companies) and the remaining 50 per cent were joint ventures. Total production of these 236 establishments was US$657 million in 1974. Electronics industry (52 factories) ranked first with a total output of US$278 million, followed by textile and garment industries (59 factories) with a total output of US$122 million. The third most important was watch and clock manufacturing (17 factories), which produced goods worth US$54 million. Taken as a whole, about 74 per cent of the total output was exported.

Domestic and International Sub-contracting

According to the 1973 Industrial Production Census, 2,228 manufacturing establishments with over 20 workers (45 per cent of total) were sub-contracting part of the their production. Total sub-contracting payment was US$166 million, which corresponded to 3.2 per cent of total value of

Table 2
Overseas Investment in Manufacturing Industries in Hong Kong (as of 31 December, 1975)

Industry	No. of Establishments	No. of Employees	Investment excluding HK Interest (in US$ mil.)	(in %)
Electronic & Electrical Products	75	27,993	138.6	(40.4)
Textiles and Clothing	71	12,372	51.2	(15.0)
Watches/Clocks & Accessories	20	5,391	38.1	(11.1)
Chemical Products	11	639	19.5	(5.7)
Printing & Publishing	8	1,182	12.4	(3.6)
Food	6	1,533	12.1	(3.5)
Toys	8	4,412	11.8	(3.4)
Building & Construction Materials	4	333	10.5	(2.9)
Metal Products	12	1,434	10.1	(2.9)
Metal Rolling, Extrusion and Fabrication	5	889	9.5	(2.8)
Others	51	3,429	29.2	(8.5)
Total	271	59,607	343.0	(100.0)

Note: 1. *Not including 1 company known to have established but not yet in operation and 19 companies known to have overseas interests but particulars of which are not yet available.*
2. *HK$4.94 = US$1.00.*

Source: Commerce and Industry Department.

production (Table 3), excluding sub-contracting by commercial houses.

Although no official statistics are available, international sub-contracting is widely practised in Hong Kong.

As shown earlier, manufacturing industries in Hong Kong are export oriented: about 70 per cent of the total output is exported. Most of the manufacturers produce according to orders received from overseas customers, and it is not uncommon for a locally manufactured product to bear an

Table 3

Sub-contracting in Manufacturing in Hong Kong (1973)[1]

Industry	No. of Establishments	No. of Sub-contracting Establishments	Gross Output (US$ Mil.)[2]	Sub-contracting Payment (US$'000)	Sub-contracting payment as % of Gross Output
Textiles	834	489	1,612	67,391	4.18
Clothing	1,250	510	1,138	44,450	3.91
Electronic and Electrical Products	375	129	598	6,582	1.10
Light Metal Products	500	284	385	13,060	6.59
Plastic Products	762	328	385	10,895	2.83
Food	132	–	209	–	–
Printing & Publishing	204	145	126	6,755	5.34
Transport equipment (including shipbuilding)	43	24	113	5,800	5.12
Photographic and Optical Goods	82	49	73	2,314	3.17
Chemicals	74	4	67	53	0.08
Paper & Paper Products	96	33	62	346	0.56
Non-electrical Machinery	77	42	41	1,153	2.80
Wood & Cork	56	16	37	746	2.01
Rubber Products	60	23	30	330	1.10
Furniture	56	21	26	1,925	7.32
Footwear	46	18	21	237	1.12
Leather	37	13	18	128	0.71
Others	281	100	224	4,023	1.79
Total	4,965	2,228	5,165	166,188	3.22

Note: [1] *Establishments with over 20 workers only. The figures do not cover sub-contracting by commercial houses.*
[2] *HK$5.08 = US$1.00*

Source: Industrial Production Census 1973.

overseas company's brand name.

The extensive practice of international sub-contracting in Hong Kong can also be known from the fact that a large number of the joint ventures identified in the C. and I. Dept. survey reported that the overseas parties provided the local firms with production and management know-how or staff to manufacture products which are marketed under the former companies' trade marks. Similar arrangements are also made under licensing arrangements.

FINDINGS OF THE SURVEY

Cross-border Commercial Sub-contracting

Commercial sub-contracting has been widely practised between developed countries (e.g. Japan, the United States and European Countries) and Hong Kong, parent companies being overseas trading houses (e.g., importers, department stores). Of the 10 companies in our sample, 8 had commercial sub-contracting arrangements with overseas firms. Increasing labour and material costs had induced the latter to sub-contract manufacturing work to Hong Kong, where such inputs are relatively cheaper. Most of the parent firms would like to sell the products under their own brand names.

Cross-border Industrial Sub-contracting

Two cases of cross-border international industrial sub-contracting were identified in our survey: one in the optical and photographic goods industry and the other in the electronics industry. In the former case, the local sub-contractor manufactures lenses and supplies them to the parent firms for use in projectors. In the latter case, the local sub-contractor manufactures electronic parts for computers according to the specifications provided by the parent firms in the United States.

In general, however, cross-border industrial sub-contracting is less common than cross-border commercial sub-contracting. The reason is that higher quality of work and stricter quality control are required for the first type of sub-contracting.

At the same time, one may safely say that, in Hong Kong, there is a tendency of increasing cross-border industrial sub-contracting, which requires a relatively higher level of technology. Because of the rising wages in Hong Kong, parent firms tend to turn to other developing countries (e.g. Republic of Korea, Republic of China (Taiwan) for sub-contracting of low technology, labour-intensive items. In order to maintain its competitive position in the world market, Hong Kong has to shift to more sophisticated products, which would require a higher level of technology, following the examples of our sample firms engaged in the manufacture of lenses and computer parts.

Within-border Industrial Sub-contracting

Within-border industrial sub-contracting is not common in Hong Kong. As there is no regulation requiring use of local parts and components, the need to build up indigenous supporting industries is not great. This may be one of the reasons why within-border industrial sub-contracting has developed little. However, there do exist some cases, e.g., electronic parts for television sets.

Table 4
Merits of International Sub-contracting for Parent Firms: Sub-contractors' View

Merits of International Sub-contracting in Hong Kong	Industry: Precision Equipment		Industry: Electrical & Electronic Appliances					Industry: Garments		
Company Number	1	2	3	4	5	6	7	8	9	10
Saving in labour cost	x	x	x	x	x	x	x	x	x	x
Saving in material cost				x	x			x		
Diversification of product range to butter against business fluctuations			x							
Sub-contract of lower quality products to concentrate production capacity on higher quality products			x							
Easiness to market products made in Hong Kong, especially to Commonwealth Countries				x						
Skillful Workers in Hong Kong							x		x	
Higher level of Technology in Hong Kong								x		
Easiness to obtain raw materials (from nearby Asian Countries)									x	

Merits of Sub-contracting

All the ten sub-contractors in our sample believed that the most important motive for their parent firms to sub-contract to Hong Kong was to take advantage of relatively lower wages in Hong Kong. Four of them mentioned easier and cheaper supply of raw materials, and three better skill and technological standards of Hong Kong as compared with other developing countries, as additional reasons (Table 4).

Marketing advantage was the most important merit of sub-contracting as envisaged by sub-contractors (Table 5). In general, most sub-contractors enjoyed a guaranteed market. Since parent firms obtained business for the sub-contractors and the products were normally sold under their brand

Table 5

Merits of International Sub-contracting for Sub-contractors

Merits of International Sub-contracting in Hong Kong	Industry									
	Precision Equipment			Electrical & Electronic Appliances				Garments		
	(Company Number)									
	1	2	3	4	5	6	7	8	9	10
Marketing advantage										
(a) guaranteed market	x	x	x					x	x	
(b) access to business connections of parent firms	x	x								
(c) reliability on parent firms' marketing efforts			x		x	x	x			x
Access to technology and skills (including design, method of production, quality control, cost control, etc.)	x			x	x		x			x
Supply of machinery and testing equipment				x						x
Access to raw materials				x			x			x
Access to improved management skills	x									x

names, the sub-contractors saved a lot of marketing efforts.

Another major motive for sub-contractors was access to more advanced technology and skills. Through association with overseas companies, local firms could easily obtain technical knowhow on product designs, methods of production, methods of cost control and quality control, etc.

EFFECTS ON PRODUCTIVITY OF LOCAL FIRMS

Eight out of the ten companies claimed that their productivity had increased as a result of receiving sub-contracts. One of the remaining two companies reported that there had been no productivity increase. The management of the second company said that it would be difficult to measure the productivity of workers because each job differed according to specifications.

The major factors that contributed directly and indirectly to productivity increase were:—

(a) access to new technology and skills of parent firms;

(b) long-term contracts and regular flow of orders, which resulted in better production planning;

(c) regular and better training of workers;

(d) expansion in markets which resulted in fuller operation;

(e) innovation in new production processes;

(f) improvement in managerial practices.

However, as pointed out by some companies, not all productivity increase was due to assistance from parent firms. There were also some internal factors that contributed to the increase in productivity: e.g., accumulation of skills over years, and adoption of better managerial practices.

TRANSFER OF TECHNOLOGY AND SKILLS

Table 6 summarizes the major types of technical assistance provided by parent companies.

The most important type of technical assistance received by sub-contractors was related to quality control. In some cases, quality control methods had been transferred from the parent company. A few sub-contractors were receiving technical information from parent firms. Some parent firms assisted the sub-contractors in quality control more actively: in one case in the electronics industry, the manufactured products were fully inspected by engineers employed by the parent firm before delivery. Another example is a garment factory, where the parent firm kept an inspector in the sub-contractor's plant.

It appears from the survey that the transfer of technology and skills takes the following forms:—

(a) Stationing of resident technical advisors: This would be the case if the

Table 6
Major Types of Technical Assistance provided by Parent Companies

Types of Technical Assistance Provided	Precision Equipment			Electrical & Electronic Appliances				Garments		
Company Number	1	2	3	4	5	6	7	8	9	10
Production techniques and know-how	x	x		x	x					x
Quality control	x	x		x	x		x			x
Product design	x			x	x					x
Supply of machinery and equipment				x						
Supply of technical specifications		x			x		x			
Supply of raw materials				x			x			x
Supply of testing equipment		x								
Introduction of managerial techniques										x
No technical assistance received			x			x		x	x	

local firm relied heavily on the parent firm for supply of technology and skills. For example; in the case of a factory in the electrical product industry, the sub-contractor had 10 resident engineers sent by the parent firm;

(b) overseas training for technical staff: As a part of the sub-contracting, technical staff of the local firm may be sent to the parent firm for training;

(c) familiarization visits to the parent firm's plant: This is normally provided to more senior technical staff;

(d) supply of better raw materials and machinery from the parent firm;

(e) provision of designs and specifications by the parent firm.

A majority of the local sub-contractors were satisfied with the technical assistance they were receiving from their parent firms. Some even felt that they did not need further assistance, because they were advanced enough to solve the technical problems themselves. Some companies indicated that what they needed was *ad hoc* assistance to solve occasional problems. Some parent firms were not in the same industry as the local sub-contractors, being trading houses or manufacturers of other products. In such case it was found difficult for them to provide technical assistance to the local firms.

PROBLEMS IN SUB-CONTRACTING

Problems and Difficulties for Sub-contractors

The greatest problem that sub-contractors encounter on a continuing basis is the restrictive and discriminative regulations imposed on Hong Kong products by importing countries. Import quota imposed by some developed countries makes exports impossible e.g. for some local textile firms to export their products even when they receive orders from overseas customers. Some Western countries regard this country as a developed country, and the general scheme of trade preference is not granted to Hong Kong. Sometimes this makes products from Hong Kong much less competitive in the world market than those from neighbouring countries. Economic recessions abroad naturally cause hardship on the part of sub-contractors here. The world economic recession in 1974–75 seriously affected Hong Kong. Indeed, production of some local firms decreased by over 50 per cent.

It is common for sub-contractors to produce similar but somewhat differenciated products for a number of parent firms, who use different brand names. Some of the interviewed sub-contractors said that this made production and inventory control more difficult. It may be noted, however, that this is a matter of individual firms' policy decision rather than a problem of international sub-contracting *per se*. There are also other problems such as difficulty in obtaining raw materials, fluctuations of business orders, lack of technical information, etc. However, these problems are usually not too serious.

Problems and Difficulties for Parent Firms

The sub-contractor interviewed during the survey indicated that their parent firms were facing some problems. In recent years, labour and material costs have been rising in Hong Kong to make it less attractive for foreign firms to sub-contract to Hong Kong manufacturers. Coupled with this is the steady increase in the land price. Consequently, some parent firms are now turning to other countries in South East Asia where land and labour are cheaper.

The already cited C. and I. Dept. survey in March 1975 throws some light on this question. Of the 236 factories with overseas interest, 58 (or

25 per cent) reported that they had operational problems. The major problem areas are shown in Table 7.

Table 7
Manufacturing Establishments with Overseas Investment that have Operational Problems (1975)

Problem	No. of Establishments
Difficulty in recruiting skilled workers	30
Inadequate floor space or land	20
Difficulty in obtaining raw materials	6
Inadequate qualities of sub-contractors' equipment	6
Others	7

Note: Some of the firms had more than one problem.

Source: Commerce and Industry Department.

GOVERNMENT'S ATTITUDE AND POLICY

The Government of Hong Kong recognizes the importance of promoting export-oriented industries and also participation of overseas firms in Hong Kong's manufacturing activities. Since international sub-contracting brings in orders for exports and sophisticated manufacturing and technical skills and creates employment, its promotion is among the objectives of the Government's campaign for industrial development.

Believing in a liberal industrial and trade policy, the Government is of the opinion that its role in industrial development is to provide a basic infrastructure and a favourable investment climate, as well as to discourage, through negotiation, importing countries from adopting a protective policy against imports from Hong Kong. It has been trying to do this by maintaining low rates of taxation and no duties on import and export, and guaranting freedom in profit remittance, capital repatriation and adequate supply of hard-working labour force.

Corporation tax is levied at a standard rate of 17 per cent. Provisions exist for the carry-over of losses from one year to another indefinitely for the purpose of profit tax calculation. As regards the supply of highly trained manpower, graduates of the universities, polytechnics, technical colleges and institutes, and graduates returning from abroad, generally meet the demand.

In order to up-grade the skill level of existing operatives, technicians and technologists, the Hong Kong Productivity Centre, the Hong Kong Polytechnics and some industrial organizations provide training programmes of short duration. Constant efforts are also made to maintain and improve the efficiency of basic infrastructure – communication systems, port facilities, banking, etc.

Probably the most noteworthy single policy is related to the supply of industrial sites. Land in Hong Kong is owned by the Crown and the Government sells the land leases normally by public auction. However, for the purpose of encouraging industrial development in Hong Kong, and the development of new technologically-advanced capital-intensive industries in particular, the Government sells such leases under the Users Restriction Scheme selectively for such projects that:

(a) are new to Hong Kong or lead to a significant technological upgrading of an existing industrial process;
(b) provide employment opportunities at a generally higher level of skill than currently available in Hong Kong, particularly for male workers;
(c) are, by their nature, unable to be carried on in standard multi-storeyed industrial buildings, and
(d) will provide essential raw materials for existing Hong Kong industries.

Two sites totalling 20 acres have been sold to two American firms, one for the manufacture of out-board marine engines and the other for the production of polystyrene. Other projects under consideration by the Government include: a steel mill with a modern foundry, a chemical plant, and several ship-building and ship-repairing engineering projects.

Another solution to the land problem is the development of industrial estates. The first of several industrial estates, which have been planned to provide land specifically for heavy industries, is now under construction. This estate will provide 136 acres of land entirely reclaimed from the sea and it will be possible to extend the area to 200 acres eventually.

The development of industrial towns and industrial zones are also means to solve the land problem. In Hong Kong, there are two well-establish industrial towns –Kwun Tong and Tsuen Wan/ Kwai Chung. Tuen Mun and Shatin are two new satellite towns and each of them has an industrial zone, which is now at an advanced stage of development.

These schemes are intended to encourage development of basic industries, which will supply essential materials to local industries and lessen the adverse effects of fluctuation in the supply of raw materials.

The most important contribution of the Government to the development of international business seems, however, to be related to the political climate: industrialists interviewed during the survey felt that the excellent political stability in Hong Kong was a factor which has been encouraging overseas companies to establish subsidiaries and joint-venture firms in Hong Kong.

Apart from these and other governmental efforts, there are also a number of programmes with which industrialists endeavour to encourage international sub-contracting. The most important is related to quality control. There is no specific legislation in Hong Kong which requires local manufacturers to meet specific technical or quality standards other than those designed to protect their users from danger. However, the Federation of Hong Kong Industries, which is a member of the International Standardization Organization, maintains a Standards Centre and provides a consultancy and advisory service on the standards required in various countries. It maintains testing and certification services for products such as textiles, footwear, plastic products, electrical and electronic goods, etc. Tests are carried out either to buyers' specifications or to standards of international organizations. The Federation also carries out pre-shipment inspections of products upon request and assists manufacturers in the quality control of their products. Industrialists are encouraged to utilize the testing facilities of the Federation.

FUTURE PROSPECTS

Thanks to these efforts, a majority of the local industrialists expect a further development of international sub-contracting in and to Hong Kong. As long as Hong Kong remains competitive in respect of labour cost, there would be an incentive for overseas companies to sub-contract their work to local manufacturers, particularly as regards higher quality products. Although wages are lower in some other countries such as Republic of China (Taiwan) and Republic of Korea, this can be more than off-set by maintaining higher standards of workers' skills and technology.

SECTION II: CASE STUDIES

(a) A Firm in the Garments Industry

The local firm, established in 1969, is mainly engaged in the manufacture of jeans, of which 92% is exported to the U.S.A., West Germany and other European Countries. The turnover of the firm in 1975 was valued at US$5 million. At the time of establishment, the company required an overseas brand name for its products in order to get access to foreign markets and contacts were made with the parent company, which is a large manufacturing establishment of fashion garments in the United States. The parent company holds 41% of the capital investment and there is no managerial or technological tie. Export to U.S. market must be made through the parent company, but in all other cases the local company is free to export using the brand name of the parent.

The lower labour and material costs and the high technological advancement in garment making in Hong Kong have been the motivation factors for the parent firm to undertake sub-contracting. For the local firm, apart from guaranteed U.S. market, it has been easier to promote sales abroad using the established foreign brand name.

No particular assistance is rendered by the parent firm except in the supply of design and information concerning the fashion trend. New designs are mailed to the local company and regular telex communication keeps the local company informed of the market situation. For the local firm, the productivity of workers has shown an increase due to the regular flow of orders and the more efficient production planning and control.

(b) A Firm in the Watches and Clocks Industry

The local firm manufactures a wide range of watches and clocks, including table alarm clocks and travellers' clocks. Its production in 1975 was valued at U.S. $8.4 million and 90% of the output was exported to U.S.A., Europe and Africa. The local firm has a subsidiary factory in Taiwan.

The relationship between the local firm and parent firm takes the form of a joint venture. Apart from the capital tie in which each party has contributed 50% of the capital, there is no other formal contractual relationship. Even prior to the establishment of the present joint venture in 1964, the local firm was engaged in the manufacture of watches and clocks. In 1964, representatives of the parent firm, who were visiting Hong Kong to locate a joint venture partner, approached the company. The products bear the brand name of the parent company, which is a large concern in the U.S.A., engaged in the manufacture of watches, clocks and also other products.

Saving in labour cost, which accounts for a substantial portion of the production cost of watches and clocks, has been the motivating factor for the parent firm. Apart from enjoying a guaranteed U.S. market, the local firm could receive valuable assistance in developing exports to other markets including U.K. Further, it has access to new techniques and skills and the parent firms provide technology and management skills in the areas of product design, innovation on production method and efficient cost control. In the initial period (2 to 3 year after the joint venture arrangement), the parent company rendered assistance in knowhow and production techniques, such as industrial engineering and assembly line layout, in quality control, product design and supply of machinery and equipment. At present, there is no particular assistance provided, since the sub-contractor is already well established. The transfer of technology and skills has been through resident technical advisers and regular as well as ad hoc visits of engineers and experts from overseas.

According to the local firm, access to new production techniques and knowhow, innovation in production processes, better training of workers and regular flow of orders have contributed to increase in productivity. In general, the firm feels that it has received adequate technical assistance from the parent firm. However, it would welcome more technical assistance, particularly in the field of research and development. According to the sub-contractor, the parent firm does not provide more technical assistance because it is afraid that by doing so, the local firm would became too technologically advanced and may compete with it.

(c) A Firm in the Optical and Photographic Goods Industry

The local firm established in 1956 and fully owned by local interests is the largest optical and photographic goods manufacturer in Hong Kong. Production in 1975 amounted to U.S.$ 30 million and all products are exported. The major markets are the U.S.A. and Europe. Under sub-contracting arrangements, it produces products under various brand names specified by the overseas companies. These parent firms include manufacturers, importers and chains of department stores. In the U.S.A., there are about 10 large buyers. In Europe, the buyers are smaller in terms of purchase quantity and there are over 100 buyers. The local company has set up sales offices in the U.S.A. and Europe.

The motivation for the parent firms has been the cheaper production costs. In general, the parent units (if they are manufacturers) would produce high quality lenses in their home countries and sub-contract the medium and low priced lenses to Hong Kong. The local firm finds that it need not be concerned with the promotional work, as this is totally handled by overseas buyers, who already have established their reputation.

The local firm has no contract on a long term or permanent basis with parent firms. However, the latter tend to place regular orders. Technical assistance provided by parent firms is very irregular and is often limited to advice and recommendations for improvement. The local firm, which initially imported the technology from Japan, has acquired necessary skill and experience over the years. If required, it would employ overseas engineers and technicians for technical advice and assistance.

(d) A Firm in the Electronic Industry

The local firm, a manufacturers of electronic computer parts, was established in 1967 as a 100% U.S. subsidiary and it employs directly nearly 450 workers. The parent is a large marketing firm in the United States, but it also handles some engineering work. The savings in labour cost and the skill and competence of the local workers in handling delicate jobs motivated the parent firm to set up the unit in Hong Kong. Staff were sent to Hong Kong from the parent firm to set up the company.

The parts are tailored-made according to the specifications of the large U.S. computer manufacturers. There are about 30 customers and 10 can be considered as major buyers. Each order from customer has a contract, which lasts from 6 months to 2 years. The parent firm finances the local firm and also supplies raw materials, designs and specifications. About 95% of the raw materials required are not available locally. In addition, the parent firm assists the local firm in quality control. The manufactured parts are sent to parent firms, where they are fully inspected before delivery.

The local firm sends 6 to 7 of its staff to the U.S.A. a year to receive training, lasting from 2 to 8 weeks. Engineers from parent firm come to Hong Kong to provide services twice a year and as required. The management of the local firm considers that sufficient technical assistance has been provided by parent firm and customers. However, it is difficult to judge the enhancement in productivity, since each job differs, according to customers' specifications.

Chapter 8

Sub-contracting in India

S.K. Subramanian

SECTION I: GENERAL REVIEW

INTRODUCTION

The growing need for foreign exchange in developing countries warrants a dynamic export strategy and a simultaneous attack on all fronts. Often, however, developing countries do not know exactly what the buyers abroad want: most of the exporters need guidance in regard to the markets, types of good and other related matters. In addition, real strength and efficiency will accrue to indigenous firms, when they go out of the protected home ground and compete in overseas markets. From these viewpoints, international sub-contracting has a special scope for contribution. This seems true specially in India with its large home market, wide-spread industrial base and, in particular, highly educated scientific and technical manpower. The presence of a number of joint-venture firms, which were set up in various sectors by leading firms from developed countries many decades ago, should also be a positive factor for international sub-contracting.

Efforts to promote sub-contracting within the country have been in vogue since 1960. With a view to developing new entrepreneurship and small and medium industries, the Government has been encouraging creation of ancillary units. Most of the public sector undertakings set examples and later it was made obligatory on all the large enterprises to sub-contract suitable items to small and medium firms. The governmental restrictions on the expansion of large firms, the credit squeeze,and the inherent large costs have been forcing these big firms to resort to sub-contracting. It is said that there are now over 10,000 units supplying ancillary items to more than 200 large enterprises, and that during 1970 – 71, the value of parts and components produced by the ancillary units amounted to nearly $50 million.

Experience of India in cross-border international sub-contracting has not been very significant, but the situation has been changing over the last few years. Initially, the huge internal demand was a hinderance to export: when a bicycle manufacturer could sell a sufficient amount of his product in the domestic market, there seemed to be no particular reason why he would bother about finding overseas buyers. But, with the growing pressure from the government to export and with the increasing confidence of Indian entrepreneurs, it has been increasingly realised that the large internal market could be a cushion against the shock that may be caused by losses suffered in export markets. On the other hand, export markets could be an insurance against recession in the home market. For example, some of the Indian auto-ancillary units have been surviving the current recession, thanks to the external demand.

No published statistics are available on the quantum of international sub-contracting, either across the border or within the country. Some indications of its rising trend are noticeable from the results of our survey, particularly in regard to cross-border sub-contracting. The survey was initiated by writing to nearly 300 companies selected in reference to literature, market reputation, and suggestions by chambers of commerce, trade associations and export promotion organisations. Nearly 100 companies replied, out of which 41 were chosen finally: seven in garment industry, one in sports good industry, eight in electronics industry, 13 in transport equipment manufacturing, and 12 in machinery manufacturing. Out of the total, 25 were foreign subsidiaries and 16 were local companies. Of the 41, 23 could be classified to be "large enterprises", holding equity of more than $1.25 million. Foreign subsidiaries' dominance in the large scale sector is notable (19 out of 23). 11 out of 12 machinery manufacturers in our sample were foreign subsidiaries. Such a relationship between the size and the ownership is not peculiar to our sample, but is characteristic of the industrialisation in the private sector of the Indian economy.

Most of the firms were engaged in more than one type of sub-contracting, and some of them were involved in international sub-contracting in many ways. For example, Lucas TVS was receiving international sub-contracts from its principal firm and other firms abroad to manufacture parts and components. Within the country, it was not only supplying the local market directly, but also receiving sub-contracts from other foreign subsidiaries and Indian firms. To meet all these orders, Lucas-TVS was resorting to both own production and local sub-contracting.

Garments

The substantial increases in the export of garments is of recent origin. It grew from only around $11.3 million in 1970 to over $200 million in 1975. The popularity of 'oriental look' or 'Indian look' in the West has been responsible for this increase. Most of the export trade was initiated by friends and relatives of some Indians who lived abroad, above all in the United States, the United Kingdom, France, and West Germany. A majority of the garment exporters in India are, aware of the fact that the 'ethnic look' is becoming out of mode and that they will have to adapt their designs to the taste of Western consumers. However, workers' training and capital facilities in this country are not good enough for this task. Therefore, the industry has become increasingly aware of the need to work as international sub-contractors with materials supplied by foreign parent firms. Some of them have already made such arrangements with western buyers and designers. For example, a firm in Delhi has Finnish designers, who use Indian fabric for the Australian market. It has also supplied, lately, garments to Sears & Roebuck, Blooming Dale, etc. Another Delhi-based firm undertook stitching of trousers for a West German parent firm using pre-cut and pre-styled fabric supplied by the latter.

During 1971, a UNIDO-Macy team visited 60 firms in India selected by the Trade Development Authority (TDA) and advised them about designs, product requirements, and sewing and finishing. As a result, substantial orders were placed. A number of delegations representing department stores, importers and mail-order houses in the United States, Canada, and Western Europe visited India during 1974–75 at the invitation of the TDA. They felt that Indian firms were capable of manufacturing any kind of garment with proper guidance pertaining to styles and sizes. Buyers from Belgium, France, and Switzerland have shown keen interest in placing orders for garments and, in particular, "T" shirts and cotton knitwear.

Recently there have been some other developments. In 1975, the TDA negotiated the transfer of a plant from West Germany to manufacture garments. The West German firm would buy 75 per cent of the output ($6.5 million) over a period of 5 years. It has also the option of purchasing the entire production. The plant is ready for shipment. In another case, an Indian firm will import equipment of $0.08 million and export 60 per cent of its production, or $0.12 million a year. Apart from these developments, the government is contemplating the creation of a "free trade zone" near Delhi to manufacture garments.

The quantum of international sub-contracting in the garment sector is thus bound to increase in the near future.

Electronics

In the electronic industry, India has plenty of human resources: professional engineers as well as skilled craftsmen. Between 1965 and 1975, the production of electronic components grew from $9 million to $102 million and that of equipment from $49 million to $300 million. Electronics exports during 1974–79 are expected to grow much faster than hitherto, due to the setting up of a diversified export base and the currently advancing export-oriented projects, e.g., export processing zones in Santa Cruz and other places. An "Electronic Trade and Technology Development Corporation" (ETTDC) has been set up recently to handle the trade between India and the East European countries.

Various surveys have revealed good market potentials for electronic components like crystal diodes, semi-conductors, capacitors, resistors, electrical insulators, carbon film resistors, wire-wound resistors, electronic valves etc., in the United States, France, Canada and Australia. Through a buyer-seller meeting in London in 1974, a considerable amount of orders were booked. Buyers from West Germany, Greece, Hong Kong, Italy and Iran have also been in contact with the TDA to locate Indian manufacturers. The TDA negotiated a deal for a State Corporation which would manufacture and export electronic parts and components under sub-contracting arrangements with a Belgian firm. The latter firm would provide necessary materials. The total export earnings in the form of labour charges will be $1.35 million over 5 years. Among other examples are the manufacture of semi-conductors for a Hong Kong manufacturer and silver-mica capacitors for a US firm. Thus, electronics in India may prove to be a fertile ground for international sub-contracting in the years to come.

Auto parts

In India, there are nearly 200 auto-parts manufacturers who fabricate such items as pistons, carburetors, radiators and shock-absorbers. For each item, there are three or more producers.

There seems to be a good chance of success for international sub-contracting in regard to items related to trucks, trailers, etc. whose markets are relatively small in any country. India could also meet other low volume demands such as spare parts of old models of vehicles, which have become obsolete. The All India Automobile and Ancillary Industries Association has done considerable spade work for developing export markets and popularizing Indian products. It has sponsored tours abroad to study the market and participated in international exhibitions. It has also produced, and exhibited, a documentary film and two publications, which depict the progress made by the Indian automobile ancillary manufacturers. The Export Inspection Council takes care of quality control and pre-shipment inspection. As a consequence of these and other efforts, the overseas demand for Indian automobile parts has been increasing steadily in Belgium, West Germany, the United Kingdom, and the United States, as well as in developing countries. Export of Indian auto-parts to East European countries amounted to $28 million in 1974–75.

International sub-contracting for auto parts is on the increase. For example, the Head of Central Materials Procurement Coordination Department of DAIMLER BENZ of Germany recently visited various ancillary units in India, and has decided to procure, for the first time, assembly parts from India. The orders executed so far, and those under process, relate to high tensile fasteners, engine mountings, oil seals, casting and forgings. The total orders would be around $0.06 million in 1976. The TDA also discussed with Klockner Humboldt Deutz, one of the most important manufacturers of tractors and farm machinery in Europe, for the supply of fuel injection pumps and forgings. Another encouraging development is the recent start of original equipment export to Australia. As a result of the visit of a delegation in February 1976, such exports would increase to $0.1 million in 1976 and the volume is expected to reach $24 million within the next five years.

These examples seem to indicate a growing opportunity of receiving international sub-contracts for Indian auto-ancillary units.

Bicycles

In addition to the 12 units manufacturing complete bicycles and 16 units manufacturing components in the organized sector, there are some 150 small manufacturing and assembling units and over 1,000 producers of parts and accessories. Export earnings rose from $0.14 million in 1960–61 to $17 million in 1973–74. Apart from the conventional Indian markets in the Middle East and Africa, Indian bicycles sell in sophisticated markets such as Canada, the United Kingdom, and the United States. Canadian firms are interested to sub-contract 3/5/10 speed bicycles with coaster hubs. There has been a growing overseas interest in bicycle components, too. In 1974, bicycle caliper brakes were introduced in the US market by an Indian firm. Another Indian firm has convinced a German firm of the superiority of its electro-galvanising techniques and the latter firm has agreed to procure a large amount of components. The potential of the market is estimated to be $18 million. France has shown interest in cycle chains and industrial chains in roll of 500 feet without connecting links. With modernisation of the component manufacturing facilities and adaptation of designs to the taste in the industrialized countries, there could be substantial scope of international sub-contracting in this area.

Hand-tools and Machine-tools

India has now attained near self-sufficiency in hand tools. Most of the large enterprises in this industry have been set up with foreign collaboration. Their high standard of technology and wide range of production, coupled with the low cost of labour, enable the industry to offer hand-tools of good quality at competitive prices. The recent boom in the 'do-it-yourself' trade in industrialised countries boosted export of hand tools from India. US auto-manufacturers also provide a good market by procuring hand tools for their "tool kits". The US buyers insist on high quality chrome vanadium tools, while demand in Germany is for high carbon tools of medium quality, where India is more competitive.

The prospects are bright for the export of machine tools of less sophisticated nature, where the advanced countries are stopping production owing to decreasing demand. IMATEC of West Germany has identified drilling machines, milling machines and metal plating machines as potential items. They have also offered help to some Indian firms to increase the precision and productivity. Canadian firms are interested in acquiring twist drills from India on a sub-contracting basis. Good export potential exists also for other items like spindles, boring bars, gear cutters, and lathe-chuckers, to be assembled into finished products.

One major problem is related to packing. US department stores and hardware stores insist on blister packing on attractively printed packing boards for display purposes. Consignment packing is another problem. The US buyers also insist on quick delivery and the long lead from India is an inhibiting factor. Shipments by containers are preferred. If Indian hand tools are to make a big dent in the world market, a consortium approach would be inevitable, since competition among Indian firms causes price erosion.

MOTIVATION OF INTERNATIONAL SUB-CONTRACTING

Motivation of cross-border sub-contracting is not simple: It is a mixture of various considerations. For example, IBM (India) exports key punch machine 029 and key entry and verifier 129 in order to get import licenses for maintaining its Indian operations. A similar example is the export of voltage level indicators to its principal firm by Philips (India). In both cases, Japan used to be the supplier. The rising labour costs in Japan may have been part of the reason why these companies started production of these items in India. Ingersoll-Rand (I.R.) exports 25 and 50 horse power compressors to the principal firm and its subsidiary in Italy to meet the export commitment, which is as high as 60 per cent of its total output. Since this became a matter of survival for the Indian subsidiary, and since the manufacturing cost is lower in India, the principal firm stopped manufacturing these two items and off loaded them to the Indian subsidiary. Similar is the motivation of GEC (UK)'s procuring fans from GEC (India) to meet demand in third countries.

The impact of these obligations and incentives is clear from findings of the survey conducted by the Indo-German Chamber of Commerce in 1973. Out of over 100 Indo-German joint-venture firms in the chemical, pharmaceutical and engineering industries, more than two-thirds of the firms mentioned governmental export obligations and incentives as their motive of export (Table 1).

Table 1: Motives of Export by Indo-German Joint-Venture Firms in India, 1973

(in % of total replies)

	Reasons for export	Percentage
a)	To avail of import entitlements	39.0
b)	To fulfil export obligations	24.0
c)	Exports are profitable	6.5
d)	Home demand is insufficient	13.4
e)	Others	
	1) to earn foreign exchange for the country	10.9
	2) to get export incentives	2.2
	3) to get sanction for renewal of the collaboration agreements	2.0
	4) to enter into new markets	2.0

Source: Survey conducted by the Indo-German Chamber of Commerce, India, 1973.

Economic calculations can also be a dominant motive. For example, Asea (Sweden) depends on its Indian subsidiary for the supply of relay panels, which is a component of a complete power system package. Asea (Sweden) finds it more economical to get these fabricated in India. The export of powerline communication carrier equipment by Hindustan Brown Boveri to its principal firm in Switzerland is also of this type. Cost calculation as well as the existence of excess capacity with Indian Oxygen induced its principal firm to sub-contract the gas regulator components for hospital equipment. This also enabled British Oxygen to utilise its capacity for diversification. Cost inflation in the UK has forced J. Stone (UK) to sub-contract to the Indian subsidiary certain items, which were sub-contracted in the U.K. previously. Purely economic considerations appear to underlie the sub-contracting to the Indian subsidiary by International Computors, Murphy, Sandvik (Sweden), etc., too.

Within-border sub-contracting is to a great extent induced by governmental directives and difficulties faced by larger units for expansion, as we already discussed in the first two pages of this chapter.

TECHNOLOGY TRANSFER FROM PARENT FIRMS

The quality of goods and the reliability in delivery are the most important criteria in selecting international sub-contractors. Even price is often of secondary importance. Most of the parent firms, particularly foreign subsidiaries, try very hard to find technically qualified entrepreneurs as their sub-contractors. In some cases, technically reliable employees are encouraged to leave the firm and become a sub-contractor.At the same time, most of the Indian sub-contractors receive technical assistance in one form or another from their foreign parent firms. Apart from provision of specifications and designs, and quality control tools, various kinds of technical advices are given by the staff of the parent firms who visit the sub-contractors frequently. The extent of transfer of technology and skills in international sub-contracting within or across the borders depends directly on the difference in technological potential between the two parties and the absorptive capacity of the sub-contractor. Broadly speaking, it seems safe to say that more efforts are made to develop sub-contractors in within-border than in cross-border sub-contracting.

i) Technology Transfer in Cross-border Sub-contracting

Garments

Out of seven garment manufacturers, in our sample, four firms depend solely on local sub-contractors, two units solely on own production and the other two on both, to meet the overseas demand.

As regards transfer of technical skills from overseas buyers, positive indications are evident in all but two cases. For example, India Exports Limited, exporter of fashion garments, totally depends on designs supplied by its overseas parent firm, which caters to the US market. The American company was formed by a brother of the Indian entrepreneur with some Americans. At first it advised only on designs, but now it also looks into the materials. For transfer of design skills there is frequent exchange of experts between the two firms. In the case of Exportos India, which is discussed in one of the section of case studies, the foreign parent firms send their designs as well as their designers to work with the Indian counterparts. This has been helping the Indian firm considerably in acquiring design skills.

Electronics

To cater to the overseas demand, three out of four sample firms resort to internal sub-contracting. In the case of two Indian firms, there is clear evidence of transfer of skills from the overseas clients. In the case of Hegde and Golay, which exports printed circuits to European Countries, Indian personnel have been trained in Europe, and the designs and specifications have been supplied by the overseas parent firms. The Indian firm had to make certain modifications in consultation with the overseas buyers, in order to use the semi-automatic and other non-sophisticated machinery available in India. In the case of Rao Insulating Company, which exports insulating components, the foreign parent firms provide designs and specifications and help for making precision dies. Experts from the clients also visit India to render technical help and for inspection.

Transport equipment

Three foreign subsidiaries and one Indian firm resort to both own production and local sub-contracting to meet the export demand and four Indian companies do all the work themselves. However, transfer of technical skills from the overseas parent firms was not evident in this group, most of the sub-contractors being well-established firms.

ii) Technology Transfer in Within-border Sub-contracting

Eighty eight per cent of the foreign subsidiaries and 55 per cent of Indian firms in the sample resort to local sub-contracting. To a great extent, this difference may be explained by the facts that the firms in the first group belong to the "large" sector and those in the second group to the "medium and small" sector, and that the governmental pressure to sub-contract falls heavier on the large firms.

Sports goods

Dunlop, India has tennis and squash rackets, hockey sticks and balls produced by local sub-contractors and export them through its foreign associate, International Sports Company (ISC). To the local manufacturers, Dunlop supplies the designs, specifications, and essential raw materials, such as ash wood and beechwood. Further, experts from ISC visit the manufacturers at intervals to sort out the technical problems.

Electronics

The local sub-contractor for International Computers India (ICI) is a large public sector undertaking Bharat Electronics Ltd. (BEL), which manufactures certain peripherals and diodes for the 1901–A computer system. The basic specifications are provided by ICI, and ICI engineers visit BEL to give advice. But due to the high technical standing of BEL, transfer of skills can be considered to be only marginal.

IBM, which is included among the case studies, educates its sub-contractors through seminars, and engineers of the Development Enginering Division visit the sub-contractors to render technical advice. In certain instances, IBM also provides the raw materials and laboratory facilities for testing. 60 per cent of the sub-contractors of Murphy (India) Ltd. are technical entrepreneurs, and the parent firm undertakes testing of components and sub-items. When failures in delivery and/or high rejection rates are observed, engineers in the Quality Control Unit get together with the sub-contractor to solve the problem. For example, the quality of electrolytic capacitors has been brought up to the required level in collaboration with an external consultant hired by the parent firm.

Philips (India) tries out most of the locally sub-contracted jobs at its factories before farming them out. Three of its research and development laboratories assist small scale sub-contractors in product development and testing and evaluating proto-types sent by them. In the case of special know-how, arrangements are made for experts to come from abroad. Philips assists the sub-contractors by ensuring regular contracts, and providing short term ad hoc loans, separate payment for tooling (occasionally given in advance) and training in aspects of production and finance. In the case of a sophisticated component like single-sided printed circuit boards, Philips provided the knowhow to an Indian sub-contractor and trained its supervisors and technicians. The sub-contractor passed on the technology, training and experience to its ancillary units.

Transport equipment

All the sub-contractors in the sample take jobs not only from foreign subsidiaries but also from other Indian firms, and all of them have direct sales in the internal market. The foreign subsidiaries (five) do not need any technical help. Among the five Indian sub-contractors, only one stated that he had not received any technical help from the parent unit, while another enterpreneur stated that he had improved his skills through his own visits abroad and through consultants.

In the case of Automac, technical guidance and help have been received from India Piston, a sister concern of Lucas-TVS, for the manufacture of cylinder liners. It is interesting, however, that the technical skills accumulated with the Automac group of industries is said to be mainly due to the mobility of personnel from large foreign subsidiaries, for which they now work as sub-contractors.

Four out of five parent firms in the sample claimed credit for imparting necessary training and assistance. The cases of Ashok Leyland and MICO are discussed later in their case studies. Lucas TVS, which has some 300 sub-contractors in windings, heat treatment, machining, moulding and forging, helps the sub-contractors in the selection of materials, quality control and in-plant training. It has helped to establish manufacturing and testing facilities as well as a heat treatment laboratory. In one case, it had helped to replace brass by aluminium in the manufacture of machined components.

Schraeder Scovill Duncan (SSD) maintains two product developmental engineers, who visit the various sub-contractors to ensure adequate quality control. Apart from the supply of designs and specifications, the SSD engineers render technical assistance, when a sub-contractor experiences difficulty. It is said that it takes 1 to 6 months to develop a reasonably reliable sub-contractor. The parent firm maintains a rating system (1-2-3) of the sub-contractors according to their technical standards. A major proportion of the work goes to the sub-contractors in the first category, despite the fact that their prices may be somewhat higher. When a shift takes place in the rating of a sub-contractor, SSD engineers diagnose the cause. The primary cause of such change has been the movement of an experienced employee of the sub-contractor.

Machinery

Regarding the development of sub-contractors, two diverse experiences were revealed in this industry. On one hand, KSB Pumps, had trained a sub-contractor in a specialised job for the manufacture of winding wires and flat cables for submersible pumps, as is discussed in one of the case studies. On the other hand, another foreign subsidiary abandoned a sub-contractor of ten years' standing in drawing and copper coating of wires, lest that the sub-contractor should become a competitor. It has started developing a new sub-contractor.

Most parent firms help their sub-contractors in developing products and procuring special materials. For example, English Electric supplies critical materials like berellium and phosphor bronze to the sub-contractors. Problems concerning manufacturing methods and material specifications have been sorted out by inviting engineers from the British principal firm. It also provides necessary tools. Hindustan Brown Boveri organizes discussions with its technical experts regularly to solve sub-contractors' problems. It also tests the sample, and when the rejection rate is over 20 per cent, Boveri feeds back the information and, if necessary, sends its engineers to the sub-contractor to set right the production techniques.

PROBLEMS OF INTERNATIONAL SUB-CONTRACTING

Probably the most serious problem is related to the supply of raw materials. To take garment industry for example, the costs of fabric alone are much more than the prices of garments which can be realized in export markets. Even in 'bonded house' operation, considerable amount of working capital gets locked up because of the central excise tax, which is not quickly refunded against exports. Another problem of materials is related to colour shade differences in Indian handloom fabric, making it necessary e.g. to take all parts of a shirt from adjacent parts of the same piece of fabric. A large portion of cutting in India is done by hand, whereas other countries use automatic machines and have even gone into laser cutting. On the other hand, the width of Indian fabrics (28 inches, instead of the standard 45 inches) makes automatic feeding of the fabric to cutting machines difficult, giving rise to high wastage ratios. Productivity of labour seems to be lower in India than in some of the competing countries, too.

The Indian automobile ancillary industry has certain inherent handicaps. As the local vehicle assemblers are small in size, they are not geared to supply large quantities. Partly for this reason, Indian components are often relatively more expensive and are not competitive in international market. Further, very few of them are equipped to produce parts of American or French designs. Some of the foreign collaboration ventures debar the Indian party from free export.

One of the most common complaints about the local sub-contractors is related to the lack of understanding of the needs of the customer, lack of initiative to improve, and absence of quality consciousness. Unless the sub-contractor has the basic ability to appreciate and absorb new skills and initiative to improve his abilities, the transfer process would become that of mere spoon feeding and it would remain inefficient.

Equally common are complaints about the lack of understanding of commercial practices on the part of the Government and about the red tape. Overseas partners often need a decision across the table, whereas the Government's decision takes three to four months, if not longer. Some reputed Indian firms could not keep the delivery date in the past because of restrictions on the import and use of certain materials like copper. Most of the foreign trade is built on mutual confidence and it is not always possible to satisfy the Government with all kinds of written stipulations and guarantees from the overseas buyers.

Complaints about the discrimination against cross-border sub-contracting and in favour of 'off-the-shelf' export appear to be valid. Under the present income tax law, 150 per cent of the export promotion expenditure can be written off from profits in the case of 'off-the-shelf' export, while in the case of exports through sub-contracting, no such allowances are made, the Government's view being that no market promotion is involved.

Locating a good sub-contractor is another major problem. One firm stated that it takes nearly 6 months to find a suitable sub-contractor.

The high freight rates, lack of adequate shipping space, limited frequency of services between certain ports, as well as restrictive trade policies of developed countries (e.g., the U.S. import quota on handloom garments and similar steps taken by Sweden and other countries) work no less against cross-border sub-contracting than against export in general.

GOVERNMENT'S POLICY

Export Incentives and Obligations

The total package of incentives include cash incentives, import entitlements, and duty draw-backs, the amounts of which vary from industry to industry and from time to time. In addition to simplifying the procedures for cash assistance and duty draw-backs, the Government has recently made all firms exporting more than 20 per cent of the production eligible for import licences for necessary inputs. Without waiting for definite export orders, advance licences can be issued for definite export plans. Many of the machinery manufacturers have been allowed to grow 5 per cent per year, solely to meet export demand. There is also a special export incentive for foreign companies. According to the Foreign Exchange Regulation Act (FERA), foreign equity participation should not exceed 40 per cent but this limit may not apply to foreign companies exporting a substantial amount. In the case of totally export-oriented enterprises the Government has permitted, in principle, even 100 per cent equity holding by a foreign firm. The 1975-76 budget has abolished the excise duty on garments.

The Government also places export obligations on individual companies, especially where foreign collaboration is involved. Recently, this has been sometimes as high as 60 per cent of the company's total production.

Inducements for Local Sub-contracting

As discussed earlier, the Indian Government has been encouraging the development of sub-contracting by large manufacturers since 1960. Since 1967 it has become compulsory for them to share part of their work with smaller enterprises. At the same time, restrictions on the expansion of larger firms, financial squeeze, etc. have induced these firms to resort to local sub-contracting. As foreign subsidiaries belong mostly to the large sector, such a policy of the Government has resulted in spread of within-border international sub-contracting.

Export Processing Zones and Bonded Factories

Together with such policies on the macro-level, the government also feels the need for schemes on the levels of individual firms and industries. In Santa Cruz and other places, the Government has constructed electronic export processing zones. It is also contemplating the creation of a free trade zone near Delhi to manufacture garments for export. In certain product lines, particularly clothing, India has already concluded agreements for shifting entire plants to India from some of the industrialized countries. Often, these deals guarantee export of a major proportion of the output and, at times, even the purchase price of the plant is adjusted against the projected export as discussed already.

The Trade Development Authority

To the small and medium enterprises, the establishment of the Trade Development Authority (TDA) in 1970 has been of great help in penetrating new export markets. Of the total exports of $80 million during 1974-75 to TDA markets, the small scale sector accounted for nearly 52 per cent and the medium-sized sector for 36 per cent. The TDA's activities cover a wide spectrum – the organisation of a trade information centre with UNDP assistance; import of samples and proto-types to make suitable product adaptations; arrangement of experts' visits to TDA clients under the auspices of UN agencies and other development programmes, and visits of overseas buyers to India; participation in international trade fairs; organization of buyer-seller meetings, etc. The TDA has also helped its clients in their negotiations for shifting plants to India from industrialized countries, along with guaranteed exports.

During 1974-75, the TDA obtained the services of a UNDP expert in SLR bicycles and frame finishing process. He visited various units and advised that more attention should be paid to the de-brassing and other processes preceding the filing of bicycle frames, as well as design improvements. The visit of another expert from West Germany and the active follow-up undertaken by the Indian firm has resulted in the introduction of the cold forging process for chain-wheels and cranks, and other modifications to the designs of spokes and nipples, and painting techniques for forks and frames.

RECOMMENDATIONS

1. Such problems in the garment industry as the use of cloth of non-standard width in automatic machines, non-availability of flawless cloth in sufficient length, non-availability of high quality cotton fabrics for garment manufacturing, etc., bear testimony to the need for a comprehensive approach.

2. In some industries, e.g., clothing, a close tie-up with designers in developed countries is essential. In the present sub-contracting arrangements, the tie-up is of very limited duration, and, for a closer relationship, joint-ventures with satisfactory export commitments may have to be considered.
3. A wider publicity on the industrial growth of India and its capabilities as an international sub-contractor is necessary. In this connection, industrialists suggested that the commercial attaches in the Indian embassies abroad should be recruited from export houses or export promotion councils, instead of among normal careermen, as people from such places are better informed.
4. It would be necessary for the government to set and operate its regulations e.g. concerning various stipulations and guarantees from the overseas clients pragmatically with due consideration of commercial customs.
5. As exporters interviewed during the survey emphasized, a stable export policy would help in encouraging cross-border sub-contracting.
6. It will be worthwhile to consider the formation of a consortium of sub-contractors in related fields, which could offer requisite guarantees, assist in the procurement of capital equipment, inspection facilities, and special material, and also negotiate with the government about other pressing issues such as credits, financial incentives, freight rates, shipping facilities, etc. Overseas buyers would also find it easy to liaise with such a central body. Some industrialists even suggested during the survey that the Government should bring some pressure or give some incentive to firms to join hands and form such a consortium. This might be started with domestic sub-contracting before spreading it to international sub-contracting.
7. Most of the people met during the survey agreed that it was important to establish a suitable machinery to handle and settle overseas complaints expeditiously.
8. Apart from international agencies, the governments of developed countries could help to promote international sub-contracting by removing protectionist i.e. misapprehensions among their people and by providing incentives for the transfer of labour intensive lines of production to developing countries.

SECTION II: CASE STUDIES

(a) Exportos India Limited

The company started in 1961 as a manufacturer of advertisement-cum-gift novelties, which it still continues to produce. Export of garments commenced in 1970, when one of the sons of the family was introduced to a few dealers through his pen friend during his pleasure trip abroad. Exports of garments have expanded rapidly and now amount to $5.3 million. To-day, the firm is one of the major exporters of ladies' fashion garments, made from handloom as well as mill-made cotton cloth. Its entire production is for export and employs nearly 1,500 workers either directly, or indirectly through sub-contracting. The garments are sold under the foreign buyers' brand names.

The rapid growth of its exports has been realised by means of frequent trips abroad, advertisements in foreign journals, participation in international trade fairs, and regular study of foreign fashion and trade journals. Its foreign parent firms include Tex-Import and Renedery in France, and Santa Cruz in the United States. They send their designers to work with Exportos' designers. This helps in the transfer of design skills.

Exportos still uses pedal stitching machines, and is now planning to import automatic machines. The firm complained about the cumbersome procedures required for import of such machinery. It now gets a cash incentive of 15 per cent of f.o.b. value, replenishment licence, and draw-back of customs duty. According to this firm, the chances of increasing exports of garments from India are good. The major constraint seemed to be the quota restrictions imposed by the importing countries. The motivation of foreign buyers to place orders with the Indian firms is savings in labour cost as well as the existence of certain special fabrics which are available almost exclusively in India.

To encourage international sub-contracting, the firm suggests introduction of extra relief in income-tax for foreign exchange earners, a liberalised policy for the import of machinery, establishment of a long term export policy, and declaration of the garment industry as a cottage industry to make it eligible for easier credits and to avoid labour problems.

(b) I.B.M. (India) Limited

Starting as a marketing organisation in 1951, IBM (India) commenced manufacturing operation in 1963. The IBM Group's policies are made on a trans-regional basis. IBM (India) exports key punching machines (029), and verifiers (129). The 129, which has 75 per cent import content, is totally exported to South-East Asia, Australia, and occasionally to Europe to meet the peak demand in that region. Such exports are imperative for IBM (India) to earn the import entitlements.

At first, it sub-contracted only simple work like assembling and surface treatment to enterprises in and around Bombay. Later, with the commencement of production of 029 machines, IBM started sub-contracting in wider fields to include more complex items like sheet metal and machined parts, die castings, electrical and electronic parts, and electro- and mechanical assemblies. At the out-set, IBM experienced quality problems, and, in some cases, work had to be brought back to its own factory or it had to set up its own plant as was the case in plating and painting.

In 1972, the company launched an "Operation Typhoon" to increase the local contents in its products partly as a result of the pressure from the Government, and partly due to the phasing out of some machines and their components in the international market. The import substitution programme for the 029 machine, and the introduction of the new key punching machine 129 played a major role in increasing IBM's sub-contracting.

To find sub-contractors, IBM resorts to country-wide exhibitions, and other advertisements. As these approaches do not always work well, however, it also seeks governmental help. For potential small-scale industrialists, IBM holds seminars at its own plants to explain its needs. A majority of the sub-contractors are small-scale units, and many of them had been with IBM through the years. This has been advantageous to both parties. In selecting its sub-contractors, a team of representatives of the purchasing, engineering and quality control units of IBM visit the candidates and check on their manufacturing capacity, equipment, skills, past labour problems, etc.

Once chosen, they are evaluated on a monthly basis in regard to their performance in delivery and quality of their products, and necessary steps are taken in order to improve their performance. IBM engineers provide assistance in tooling, production processes, quality control, etc. Whenever a specialised item is to be sub-contracted, IBM trains the sub-contractors' employees at its own plant and provides necessary equipment and materials. However, if after all the efforts the quality and delivery performance does not come up to the minimum acceptable standard, the sub-contractor is eliminated. Normally the acceptable maximum rejection ratio is around 20 per cent. In the earlier stages, the sub-contractors are not particularly conscious of quality and delivery requirements. The unyielding pressure from the Purchase Department and the visits of personnel from the Quality Assurance Department help a lot in this regard. The frequent contacts made by IBM representatives from the purchasing, engineering and quality control units help to increase the mutual understanding.

In order to overcome the problem of non-availability of proper machinery, IBM sometimes develops alternative production methods for sub-contractors. For example, sheet metal parts requiring fine die-cutting are shaved or ground when fine die-cutting equipment is not available. In another case, parts which are normally finished in one operation by broaching is produced in three stages by milling and grinding. For a number of parts, IBM has more than one sub-contractor for the purpose of ensuring their satisfactory services in regard to price, quality, and stability of supplies. Three to four months' supply of locally sub-contracted items is kept in inventory. Suppliers of items with multiple sources belong to the small-scale sector and over 90 per cent of them are technical entrepreneurs.

IBM mentioned the following as major problems in sub-contracting in India; inadequate manufacturing equipment, inspection and quality control facilities, non-adherance to delivery schedules, difficulty in procuring raw materials, and lack of understanding of IBM's needs on the part of sub-contractors.

(c) Philips (India) Limited

Philips commenced its operation in India in 1930, and today it has ten factories, employing over 7,000 workers. Its annual turnover is $80 million. Sixty per cent of the equity is held by the principal company. Its marketing network consists of nearly 4,000 dealers. As a corporate policy, Philips (India) exports only those items, which can also be marketed in local market. At present, about 50 per cent of its exports are on account of sub-contracts received from abroad. The percentage is expected to rise to 60 per cent in the near future.

Philips exports parts and components (e.g. PVC gang condensors, loud speakers, voltage level indicators, carbon resistors, ceramic bodies, intercoms, multimeters, radio and tape recorders) to nearly 30 countries. Some items are exported to its principal firm. They include voltage level indicators which used to be procured in Japan. Although the device is fairly simple, it took Philip (India) nearly a year and a half to develop it. It also exports intercommunication systems to its principal firm. They have to meet international standards, and so they are exported without switches, plugs, and sockets, as these parts available in the local market do not come up to such standards.

Philips has 561 sub-contractors, of which 481 are in the small-scale sector, each having an average turnover of about $0.2 million (1974 figures); 77 are medium-sized, and 3 belong to the large-scale sector. The small-scale sub-contractors are not captive to Philips and their aggregate turnover due to this parent firm is around $14.7 million. The total raw material requirements of Philips are nearly $26.6 million: nearly 12 per cent of this is imported and nearly 55 per cent is supplied by the small-scale units. The parent firm stated that, as a matter of company policy, only technical entrepreneurs, and not ex-trader entrepreneurs, were eligible as its sub-contractors. The purchase department maintains a list of potential sub-contractors, sends out enquiries to locate new suppliers, and assesses their capability through visits. The central purchase unit comprises of four people only, but the production staff is drawn upon for assessment visits.

As a matter of policy, two sub-contractors per item are developed. Philips assists their sub-contractors by ensuring a stable contract and providing short-term loans on an ad hoc basis and separate payments for tooling. It also helps them technically in regard to selection and layout of machinery, labour management, production planning, design of tools, special manufacturing techniques, raw materials and quality control facilities. It is stated that most of the locally sub-contracted work is tried out by Philips at its factories before farming out. Three of its R & D laboratories assist the small scale sub-contractors in developing and testing of prototypes. In the case of special knowhow, arrangements are made for experts to come from abroad. Training in all aspects of production, finance, and marketing is also provided for the sub-contractors and their employees.

(d) Ashok Leyland

Leyland Motors, U.K., holds 51 per cent of the equity of this company, which commenced its production in 1956. With an investment of $17 million, its annual turnover is $50 million, of which $13 million is exported. The company manufactures 8,000 Comets 7½ and 12 ton payload trucks, and 1,000 marine engines annually. For light vehicles, it does not import any component, whereas for the heavy ones, imported components constitute 25 per cent of the total cost of components, mostly crank shafts, frames, chassis and side members.

Leyland Motors maintains a central spare parts pool, and Ashok Leyland supplies spare parts of engines, which have become obsolete in Britain. The principal firm helps in marketing the products in foreign countries. The exported trucks would bear Leyland's trade mark. Against exports, the company is entitled to cash incentives, replenishment licences and duty draw-backs.

To illustrate its role in the development of sub-contractors, Ashok Leyland referred to three instances: development of Comet side members, tappets, valve guides and crank shafts. Until two years ago, the side members of Comets were imported from the principal firm, as import prices were lower than the prices quoted by Indian companies. In the meantime, the Comet model became obsolete in Britain and the principal firm ceased to supply the parts. Ashok Leyland got in touch with Press Metal Corporation in Bombay, which had had some experience in cold rolling. Ashok Leyland got cold rolling technology from British Leyland and also helped the sub-contractor in setting the roll at the correct pitch, choosing appropriate materials, and inspection and quality control. It also arranged visits of Japanese experts to get advice pertaining to raw materials imported from Japan. In addition, a visit of technical staff of Press Metal Corporation to Ayrshire Metal (UK) for in-plant training was organised. Similarly, Ashok Leyland transferred technical skills to Indian sub-contractors for the manufacture of tappets and valve guides from special chilled cast iron. The manufacture of crank shaft by Bharat Forge was also a result of joint efforts by the two parties.

(e) Motor Industries Limited (MICO)

Robert Bosch of West Germany holds 51 per cent of the equity in MICO, which commenced production in 1953 and makes diesel fuel injection equipment, spark plugs for petrol engines, filter units and stationary small horsepower engines. With an investment of about $17 million, its annual turnover is around $57 million and its exports amount to $12 million. Export is made to Bosch and its subsidiaries and agents. 75 per cent of the exports are made to the German principal firm, who sells the products under its own brand name. A major portion of the remaining exports goes to France, Italy, Argentina and Brazil, where the products are marketed under the brand name of Bosch's subsidiary in each country. The exports to Middle East, South East Asian countries and Sri Lanka are handled by Bosch agents, but the products are sold under the brand name of MICO. Against these exports, MICO is entitled to a cash incentive of 20 per cent of the f.o.b. value, replenishment licence equivalent to 20 per cent of f.o.b. value and duty draw-back on imported materials.

MICO resorts to internal sub-contracting to the tune of $1.7 million per annum. Over 200 sub-contractors are involved. MICO has also a number of sub-contractors who supply gaskets and other pieces of leather, rubber, nylon, and plastic. In order to locate its sub-contractors, MICO receives assistance from the Small Industries Service Institutes, and visits the selected enterprises to examine their abilities and facilities before placing orders. Once a sub-contractor is chosen, all the technical knowhow, designs, dies, and special as well as general gauges for quality control are provided for him. Sometimes, entrepreneurs do not know how to use their facilities to make specific items. In such cases, MICO gives them advice and training. It also helps in costing and pricing their products. In-plant training for sub-contractors and their employees is also provided.

MICO's main motive for sub-contracting is savings in capital investment, and there is an implicit agreement with the sub-contractors that they have to supply 50 per cent of their production to MICO and that for the balance they are free to sell in open market. Ninty to ninty five per cent of the suppliers are regular sub-contractors. As a policy, MICO keeps alternative sources of supply even if the required amount is very small.

MICO's major problems with its sub-contractors are: a) shortage of skilled personnel who work on automatic machines, b) lack of quality consciousness, c) lack of initiative, e.g., to improve or to find alternatives, and c) lack of delivery consciousness.

(f) KSB Pumps

KSB Pumps manufactures centrifugal and submersible pumps. A wide variety of materials are used such as cast iron, bronze, aluminium bronze, nickel cast iron, steel and stainless steel. With an investment of $0.67 million, its annual turnover is nearly $6.5 million. Its German principal firm, which is engaged in manufacturing activities in 14 countries and owns selling agencies in 45 countries, holds 65 per cent of the equity. By contract, KSB can export finished products only to South East Asia, but no such restriction applies to the export of components. In 1974, KSB exported finished pumps worth $0.66 million to Bangladesh and South East Asian countries. It is negotiating with its principal firm for the supply of finished goods. KSB receives technical information from its principal firm free of charge. Against exports, KSB gets cash incentive of 20 percent of f.o.b. value, replenishment licence equivalent to 40 per cent of f.o.b. value, and duty draw-back on imported items.

Nearly 30 per cent of the value of the turnover is procured through within-border sub-contracting. Major sub-contracted items are castings such as pump impellers, stuffing boxes and bearings.

At a very early stage of sub-contracting, KSB learnt that foundries in India did not possess enough knowhow to produce castings required by them. KSB obtained necessary information and also the services of a top foundry specialist for a period of one and half months from its German principal firm. The expert paid visits to all the sub-contractors and told them about the methods to be adopted. He worked for one month at the factory of one of the major casting suppliers to transfer the knowhow for alloy steel and steel casting. Two of the technicians working with the sub-contractors were trained at the principal firm's foundry for a period of six months. Even now, when a technician from a sub-contractor visits Europe, he is received by the German firm and technical problems are thoroughly discussed. Many sub-contractors greatly appreciate the merits of such visits.

As a very recent example, KSB referred to the improvement in the finish of impeller castings from Modern Foundry & Machine Works, Ahmednagar, after the visit of their factory manager to Germany in May 1976. Valuable technical information has also been passed on to other foundries like Kirloskar Brothers Ltd., and Precision Tools & Castings Pvt. Ltd., Lucknow. This has helped them particularly in the manufacture of hydraulically pressure-tested castings. KSB also passed on to Kirloskar the impeller designs and patterns, which resulted in a considerable improvement in the performance of this sub-contractor's impeller.

Another case of successful transfer of technology and skills is related to winding wires and flat cables for submersible pumps. Such winding wires and cables were not manufactured in India, when KSB entered the industry. KSB obtained the samples and knowhow from its principal firm and passed them on to Finolex Cables Ltd. The endurance test of these products were conducted by KSB. By providing samples, preparing drawings, and providing necessary guidelines to achieve the required performance, KSB also helped a subcontractor to develop various products including drip type constant level oiler, which has come to be used by all the pump manufacturers in India.

Chapter 9

Sub-contracting in the Republic of Korea

Korea Productivity Center

SECTION I: GENERAL REVIEW

INTRODUCTION

The practice of international sub-contracting in the Republic of Korea began to increase substantially only in the late 1960's, when the Second Five-Year Economic Development Plan (1967-71) was successfully implemented. The Third Five-Year Economic Development Plan (1972-76), which brought about the rapid expansion of Korean export industries, has further stimulated it. In earlier years, however, international sub-contracting was confined largely to clothing and electronic industries. It was after the adoption of a new policy aiming at the development of heavy industries by the Government in the mid-1970's that similar practices started spreading in such industries as transport equipment (e.g. automobile and shipbuilding) and industrial machinery manufacturing.

This section is intended to provide general information about the development, present status, and future prospects of international sub-contracing in garments, electronics, automobile and ship building, and machinery manufacturing industries. Due to the lack of information, our discussion in this section will be confined mostly to the cross-border sub-contracting.

Garments

Manufacture of garments was the first industry to develop through international sub-contracting. The export of garments under the so-called bonded processing arrangements commenced at the end of the Korean War (1950-53), but the expansion of the sector as an export industry began in 1962. Since then the industry has grown gradually from old-fashioned household industry into a large-scale modern sector, and its products have become increasingly diversified. The number of garment exporters with more than 500 sewing machines, for example, was only slightly over 20 in 1968, but it rose to 252 in 1972 and 275 in 1974. Export earnings of these firms grew from $149.7 million in 1972 to $427.1 million in 1974.[1] The mounting labour shortage in some developed countries, the United States and Japan in particular, has contributed a lot to this rapid expansion. These firms do not produce their export products until they receive orders from foreign buyers, under whose brands the products are marketed. Each of these Korean firms is known to have on an average around ten such parent firms.

Electronics

Electronic industry has made rapid progress since 1970, when it was designated by the government as one of the strategic export industries. To accelerate its development, foreign investors were permitted to sell some of their products in the domestic market. Foreign investment became active particularly in the manufacture of vacuum tubes, semi-conductors, passive circuits, and so on, which were either used for local assembly or exported as parts.

In 1970, there were 184 electronic units, including seven foreign and ten joint-venture firms. Though small in number, these two types of firms accounted for 47.3 per cent of the total investment in the industry and exported $40.5 million worth of goods, which was equivalent to 74 per cent of the industry's total exports in the same year. By 1975, the number of electronic manufacturers had risen to 389, of which 108 and 40 were joint ventures and foreign firms, respectively. Total exports had grown more than ten times to $581 million during this period, while the share of joint-venture and foreign firms had almost remained unchanged (Table 1). This strongly indicates the importance of international sub-contracting in this industry.

Automobiles

In Korea there are four manufacturers of passenger cars, buses, trucks and other vehicles; Saehan, Hyundai, Asia, and Kia. They had been assembling these automobiles, using parts and components imported in the CKD form, until 1973. In that year, however, the Ministry of Commerce and Industry announced the "People's Car" programme, which has given a great impetus to the local production of parts and components. By the end of 1975, Korea was able to turn out 70 per cent of parts and components for passenger cars, 60 per cent for trucks, and 85 per cent for buses, in

---

[1] The figures were provided by the Garment Exporters' Association.

Table 1
Exports of Electronic Products by Type of Firms
(1970 – 1975)

(in thousand dollars)

Year	Domestic firms	Joint-ventures	Foreign firms	Total
1970	14,507 (26%)	7,591 (14%)	32,866 (60%)	54,964 (100%)
1971	24,712 (28)	11,139 (12)	52,752 (60)	77,582 (100)
1972	41,050 (29)	23,501 (17)	88,603 (54)	142,133 (100)
1973	115,515 (31)	72,573 (20)	181,204 (49)	369,292 (100)
1974	145,244 (26)	111,598 (20)	296,396 (54)	553,238 (100)
1975	152,000 (26)	128,000 (22)	301,000 (52)	581,000 (100)

Note: The figures in the parentheses indicate percentages of the totals.
Source: Korea Precision Instrument Center.

terms of the number of pieces. It is planned to produce 95 per cent of parts and components for the "People's Cars" by the end of 1976.

In 1971, there were 249 firms in this industry, which produced more than 1,500 varieties of parts and components. Exports increased from $1.2 million in 1971 to $10.2 million in 1975. In the latter year, exports of completed vehicles amounted to $0.5 million (Table 2). At least part of the export has been under international sub-contracting arrangements,e.g. those between Saehan and other members of the GM group abroad. Most of the exported parts and components are directed to the replacement of markets.

Table 2
Exports of Automobile Industry
(1971–1975)

(in thousand dollars)

	1971	1972	1973	1974	1975
Cars	–	–	–	–	495
Parts and components	1,159	2,182	5,038	8,507	9,688
Total	1,159	2,182	5,038	8,507	10,183

Source: Ministry of Commerce and Industry.

Shipbuilding

The latest development in the domain of international sub-contracting is the expansion of shipbuilding since 1972. In view of the considerable industrial linkage and employment effects of this labour-intensive industry, the Government has been encouraging it as one of the strategic export industries, with its efforts aiming at the development of local supply base of materials, designing and other skills. As a result, its output rose from 50,500 G/T to 612,500 G/T and its exports from mere 2,800 G/T to 588,400 G/T, between 1972 and 1975. It is intended to raise the production to 2.5 million G/T by 1981, with the completion of the Fourth Five-Year Economic Development Plan.

Machinery

International sub-contracting in the machinery industry in Korea is relatively less significant. This has been due to insufficient accumulation of capital and technological knowhow, under-development of related industries and the small size of the domestic market. As shown in Table 3, Korea has been importing more than 50 per cent of her machinery requirements.

Table 3
Demand for and Supply of Machinery in Korea (1970–1974)

(in million dollars)

	1970	1971	1972	1973	1974
Total demand	1,249	1,327	1,462	2,263	3,299
Domestic demand	1,229	1,236	1,282	1,845	2,387
Exports	65	90	180	418	914
Domestic production	683	613	671	1,066	1,282
Dependency on imports	47.2%	53.8%	54.1%	52.9%	61.1%

Source: Ministry of Commerce and Industry

The quantum of foreign investments in this sector was around $143 million by the end of June 1974, representing 36 per cent of the total foreign investments. Out of the 395 contracts concluded with foreign firms to import industrial technology between 1962 and 1973, only 144 of them related to the manufacture of machinery and most of the other contracts were in the fields of electronics and transport equipment. Promotion of international sub-contracting in the machinery sector would warrant further encouragement to international co-operation in this industry. A new chapter in the machinery industry was opened in 1974, with the completion of the Changwon Machinery Industrial Complex on the southern coast. Further development of the industry is expected as a result of increased foreign investments and technological tie-ups with foreign firms.

SECTION 2: CASE STUDIES

(a) Parent firms

Twenty parent firms, all located in Korea, composed of eleven in electronics, five in machinery, three in automobile and one in shipbuilding industries, were covered during the survey. These samples were chosen on the recommendation of the industrial associations. All the parent firms were sub-contracting production or processing of parts and components, and only two of them had assemblers as their sub-contractors. Being well established foreign firms or joint ventures, the parent firms covered in the survey had a large number of local sub-contractors. Some of the sub-contractors had been newly established, specially to meet the needs of the parent firms. Industry-wise and size-wise classification of the sub-contractors is shown in Table 4.

Table 4
Number of Sub-contractors by Industry and Size

	Electronics	Machinery	Automobile	Shipbuilding	Total
Large	89 (6)	2	69	30 (2)	190 (8)
Small	718 (5)	57	189	73	1,037 (5)
Total	807 (11)	59	258	103 (2)	1,227 (13)
Average per parent firm	73.4	11.7	86	103	61.4

Note: The figures in the parentheses indicate the numbers of those firms which were established newly as sub-contractors. "Small" firms are firms with fewer than 200 workers.

The quantum of purchases from the sub-contractors were, however, not very substantial except in the case of shipbuilding. Out of the total material procurements of the parent firms, supplies from sub-contractors amounted to 28.8 per cent in electronics, 18.8 per cent in machinery, 33.3 per cent in automobiles and 71.0 per cent in shipbuilding. No foreign firms in garments and bicycle industries had local sub-contractors.

(b) **Sub-contractors**

On the recommendation of parent firms and industrial associations, 51 sub-contracting firms were chosen for the survey. An industry-wise and size-wise breakdown of the samples is shown in Table 5. Fifteen of the firms were engaged in cross-border international sub-contracting with their parent firms abroad. All the five sub-contractors in the garments industry belonged to this category. Four firms in shipbuilding and electronic industry had been set up for the special purpose of working as sub-contractors.

Table 5

Breakdown of Sample Sub-contractors by Industry and Size

	Auto-mobile	Ship-building	Elec-tronics	Mac-hinery	Garments	Total
Large	4	2 (2)	13 (1)	3	–	22
Small	12	–	8 (1)	4	5	29
Total	16	2 (2)	21 (2)	7	5	51
Those which have parent firms abroad	3	–	6	1	5	15

Note: The figures in the parentheses show the numbers of firms specially established as a sub-contractor.

Excluding the garments industry, the sub-contractors, on an average, had 21.1 parent firms in Korea. This is, however, largely due to the heavy weight of "large" enterprises among the sub-contractors in electronic industry. On the average, sub-contracted work accounted for 52.9 per cent of the total sales of the sub-contractors. However, inter-industrial differences were considerable: 100 per cent in garments, 57.5 per cent in shipbuilding, 56 per cent in automobile, 51.0 per cent in electronics and 16.3 per cent in machinery. Except in the garments industry, it is, however, not very clear as to how much of the sub-contracted work was of international nature.

TWO PARTIES' RELATIONSHIPS AND MOTIVATIONS

In most cases, the parent firm and the sub-contractor found their partner in sub-contracting by direct contact. The second most important means of discovering such a partner was through introduction from another company or a friend (Table 6).

Table 6
Means of Establishing Sub-contracting Relationship

(No. of replies)

	Parent firms	Sub-contractors
Direct Contract	16	48
Introduction by another company or a friend	11	5
National fair in Korea	3	2
Govermental information services in Korea or in the parent firm's home country	2	4

Note: Some firms indicated more than one source of contact.

Table 7
Motivation of Sub-contracting

(No. of replies)

a) Parent firms	
Specialised items which the parent firm cannot produce and which are cheaper to buy locally than to import	12
Expansion of production capacity	8
Savings in labour cost	4
Savings in material cost	4
Easier access to local market	3
Restriction on import of parts and components used by the subsidiary, imposed by the Korean Government	3
b) Sub-contractors	
Marketing advantages	37
Access to new technology and skills	9
Better access to materials	4
Financial advantages	3
Other	3

Note: Some firms gave more than one reason.

Most of the parent firms had been sub-contracting to local firms for economic reasons, above all, in pursuit of lower costs. To the sub-contractors, advantages in marketing were by far the most important merit of sub-contracting, followed by the access to new technology and skills (Table 7).

It is interesting to note that, in spite of the dominance of cost factors among the parent firms' motives of sub-contracting, "low cost" was not so important a criterion in selecting their sub-contractors, as reliability in regard to quality and delivery of their products. The sub-contractors seemed to pay much attention to payment conditions, and the amount and stability of orders (Table 8).

Table 8

Criteria in Selecting the Partner

(No. of Replies)

a) Parent firms	
Standards of equipment and technology	15
Ability of quality control or low rejection ratio	14
Cost of production	11
Certainty of delivery time	12
Managerial abilities	2
b) Sub-contractors	
Conditions of payment	11
Size of paid-in capital	8
Amount of order	7
Continuity of the relationship	3
Advantages in marketing	3
Other	6

Note: Some firms indicated more than one criterion

Probably due to the relatively short history of this kind of business practice in Korea, the relationship between the two parties was reported to be irregular in a great majority of the cases (16 out of 20 parent firms). Accordingly, 16 parent firms and 45 sub-contractors in our sample said that they had no special (i.e. financial, managerial or technological) tie with their partners.

TRANSFER OF TECHNOLOGY AND SKILLS FROM PARENT FIRMS

All sub-contractors covered in the survey reported that their poductivity has improved as a consequence of receiving sub-contracts from foreign parent firms. Fuller operation of their production facilities due to increased orders and better planning of their work due to a more regular flow of orders were most predominent among the factors cited by the sub-contractors. (Table 9).

Table 9

Factors Contributing to Sub-contractors' Increased Productivity

(No. of replies)

Increased sales and consequent fuller operation	42
Regular flow of orders which permits better planning	32
Improvement in quality of products	21
Better training of workers	16
Innovations in production processes	15
Decrease in production cost	7
Introduction of new products	5
Replacement and better maintenance of obsolete facilities	4
More favourable supply of raw materials	2

Note: Some firms indicated more than one factor

As regards the mode of technical guidance and technology and skill transfer, exchange of visits by technical staff from both sides was the most common. All the 20 parent firms mentioned irregular visits to sub-contractors by their own staff, and seven of them referred to regular visits, as their methods of technical guidance. Only one of them had been training sub-contractors' staff in its own plant.

Forty two out of fifty one sub-contractors mentioned mutual visits by the technical staff from both sides as their means of transfer of technology and skills. Three of them had been receiving regular visits of technical advisers from their parent firms, three had resident technical and/or managerial advisers, and one had been having its staff trained at vocational training institutes through arrangements made by its parent firm.

Only four sub-contractors reported that they had been receiving sufficient technical and/or managerial assistance from their parent firms. These were joint-venture firms, which had close ties with their parent units. The rest of the sub-contractors covered in the survey had somehow failed to develop mutual trust and understanding with their parent firms. This, as well as inadequate governmental regulations pertaining to such assistance seemed to be the major constraints. In the opinion of the parent firms, lack of basic technology and skills, and shortage of capital, equipment and machinery and adequate raw materials on the part of sub-contractors, were regarded to be main causes of difficulty in providing sub-contractors with new technology and skills (Table 10).

Table 10
Factors Limiting Technology Transfer: the Views of Parent Firms

	(No. of replies)
Lack of basic technology and skills	10
Lack of equipment and machinery	9
Shortage of capital	4
Inadequate supply of raw materials	3
Other	4

Shortage of capital and adequate raw materials are real problems, so much so that sub-contractors wanted to have such basic requirements provided by their parent firms, rather than technology.

PROBLEMS

To the parent firms, the most common problems of sub-contracting in Korea appeared to be poor quality of raw materials, capital equipment, and workmanship in the order of seriousness, while delayed payments, unjustified cancellation of orders, and shortage of capital were the problems most commonly mentioned by the sub-contractors (Table 11).

Shortage of adequate materials is a very serious problem for both the parties. This is particularly so in electronic industry and every surveyed firm mentioned it. Shortage of capital, both working and fixed, is more acute among small enterprises than in the case of large enterprises. Accordingly, the former enterprises urged that they should be given a greater financial support by the Government. (In reality, the share of medium and small enterprises in the total amount of long-term loans fell from 29 per cent to 22 per cent between 1973 and 1975.)

Most sub-contractors faced the problem of shortage of technicians and skilled workers, the case being most serious in automobile and machinery industries. Coaching of such personnel by large enterprises had been aggravating the situation in small enterprises. Closely related to this issue was the lack of training of employees of the sub-contractors. In most cases,

Table 11
Problems of International Sub-contracting

(No. of replies)

a) Problems of Parent firms	
Poor quality of materials	10
Poor quality of equipment and machinery	6
Poor quality of workmanship	5
Unreliable delivery	4
High rejection ratio	4
Difficulty in finding suitable sub-contractors	4
Taxation problems	2
High transport costs	1
Other	3
b) Problems of sub-contractors	
Delay in Payment	37
Unjustified cancellation of orders	23
Shortage of capital	22
Inadequate materials	17
Lack of skills	15
Lack of information about available sub-contractors	13
Other	4

Note: Some firms indicated more than one problem.

they did not have any training programme conducted by adequately qualified staff, nor did they receive sufficient technical asistance from the parent firms. On the other hand, some sub-contractors pointed out that parent firms provided only samples rather than designs and blue prints, which discouraged their efforts to develop new technology and skills.

Parent firms' purchasing policy also seemed to have some relevance to this problem. Sub-contractors reported that most parent firms demanded cutdown in prices, neglecting the degree of difficulty involved in the work. The general practice was said to be a 10 per cent discount of the price quoted by the sub-contractor. By ordering the same products from several sub-contractors parent firms encouraged competition among their sub-contractors with a view to bringing down the prices. These practices were considered to be responsible for the failure in improving technical standards on the part of sub-contractors. The fact that most of the sub-contractors are producing too small amounts of too many items must also have a negative impact in this regard. Finally, it is noteworthy that the problem of unreliable delivery accrues largely from sub-contractors' lack of systematic,and scientific management methods regarding production and quality control.

CONCLUSION

International sub-contracting has not been adequately encouraged, or its merits have been underestimated, by both the government and entrepreneurs in Korea. The government has tended to regard the practice as a convenient means of inducing foreign capital investment and expanding export, while industrialists have been attempting to develop new products superficially by imitation. For this reason, the practice of international sub-contracting has made, so far, little contribution to the development of local technology and skills and the linkages with local industries have been rather limited. Parent firms and sub-contractors have remained, largely, to be mere buyers and vendors, exerting no sincere and earnest efforts to improve industrial technology and skills in Korea.

For a successful implementation of the government's programme for the development of heavy industries, co-operation between foreign parent firms and local sub-contractors must be intensified in many ways. In this respect, it is expected that the situation will improve considerably, as the government intends to strengthen small enterprises as necessary supporters of its programme for heavy industries during the Fourth Five-Year Economic Development Plan, which starts in 1977.

In order to encourage technology transfer from parent firms, however, it seems necessary to:

(i) Standardise products, and

(ii) Introduce some institutional machinery to ensure the transfer of technology and skills from parent firms.

Chapter 10

Sub - contracting in Pakistan

K.M. Riazuddin

SECTION I: GENERAL REVIEW

INTRODUCTION

The concept of sub-contracting is not new in this sub-continent. In fact, such business practice started during the Second World War. As the War continued and as supplies from abroad became difficult, the Defence Forces procured only the core components from abroad for assembly and sub-assembly of military goods, while the main contractors sub-contracted the manufacture of certain parts and components to smaller enterprises. After the War, when it was no longer necessary for the Defence Forces to rely on local industries, almost all the enterprises set up during the War were forced either to close down or to divert to production of non-military goods. The industrial sub-contracting which had thus developed during the war, more or less came to a halt.

At the time of the partition of the sub-coninent into Pakistan and India, manufacturing industry worthy of the name was all located in the territory that became India, and Pakistan was left with hardly any significant industrial base. With concerted efforts, however, new enterprises were started, specially textile mills. These mills not only fed yarn to their own looms but also met the need of hand-loom industry and surplus yarn was exported. This practice continues to this day. In other industries, every component of a product used to be manufactured and/or processed under one roof, mainly because there was hardly any ancillary industry.

It was only in the late 1950's or early 1960's that people started talking about sub-contracting. Industrialists in developed countries had by then become aware of the fact, that their wages were rising in their countries and that the day would not be far when they would have to face the problem of rising costs of production. To overcome this problem by taking advantage of low cost labour and facilities offered by developing countries, they started shifting part of their assembly-cum-production activity to such countries. It was in the early 1960's that few firms, particularly those in machinery manufacturing industry, started looking into the feasibility of joint ventures in Pakistan. About half a dozen of well-known enterprises were thus set up in the 1960's.

At the same time, need of diversification of Pakistan's industrial base, expansion of manufactured and semi-manufactured exports, employment promotion, etc. started attracting attention of policy-makers. It was all these considerations that gave rise to a serious thinking about the development of sub-contracting both at the national and the international levels. However, not much was achieved because of the political instability in the country, the war of 1971 with India, the division of the eastern and western parts of the country, and the after-effects thereof.

Nevertheless, a certain amount of sub-contracting was practised. Textile mills and other industrial units which received orders were not always in a position to fabricate everything themselves. They procured certain parts and components from other manufacturers.

A recent development is a similar practice in the manufacture of machinery and equipment, two wheelers, trucks etc., where some of the parts and components are sub-contracted to local manufacturers or bought in local market. The process of such procurement is called "trading business", or "contracting", and the term "sub-contracting" is still to be introduced.

The same is the case with the international business. Garments and some other products are made to the foreign buyers' designs either entirely in the exporters' own factories or in cooperation with their sub-contractors. Yet, they want to be termed as "contractor" and not as "sub-contractor" to a foreign parent company.

SUB-CONTRACTING IN PAKISTAN

In many ways, industrial sub-contracting in Pakistan is still at its exploratory stage. For the procurement of components, samples are received from one or more fabricators to see which is up to the required specifications or meets the requirements best. Only after a satisfactory prototype has been produced, sub-contracting arrangements are made with the selected fabricator. In most cases, the sample is tested and approved by the parent firm and/or its principal firm before regular production starts. This is necessary, particularly because raw materials of required specifications are in short supply in this country and foreign parent firms are not willing to let their name be used for or in relation to products which do not meet their minimum acceptable standards.

In any case, organised sub-contracting has just started at both national and international levels. The foreign parent firms are still not very conversant with the facilities available in Pakistan, and they at times route their enquiries to the Export Promotion Bureau of the Government of Pakistan, who advertises such enquiries for the benefit of both the parent company and the potential sub-contractors and thus helps in bringing the two parties together.

PROGRESS IN RECENT YEARS

As stated above, the number of sub-contractors have never been very large in Pakistan. Even the increasing trend in recent years has not enlarged it significantly. Most of the existing sub-contractors set up their business after the 1971 debacle and they are not very experienced in the trade. Many of them, however, are international sub-contractors: in some cases they manufacture their products according to specifications provided by their parent firms, and export them under these firms' trade marks. In other cases, they produce parts and components which are assembled and marketed in Pakistan, using their principal firms' brands. Part of such work is further sub-contracted to other local firms. Products such as ladies' clothing, sports goods, surgical instruments and cutlery are exported mostly without any brand name in order to enable the foreign retailers to sell them after imprinting their own brands.

In the case of surgical goods, two or three large manufacturers have their own offices in countries like Australia, Britain, West Germany and the United States. Others sell goods in semi-finished forms to their parent firms, which give a finishing touch to them at their own works or the works of different sub-contractors, repack and sell them under their own brands. There were, however, instances where the foreign buyer came to Pakistan personally, had the products fabricated and prepared to his own designs and specifications and then despatched to destinations of his choice.

Subsidiaries of foreign firms of repute which have production or assembly plants in Pakistan, are assisted by their principal companies, which at times share received orders with them. Such enterprises are, however, not very many. All these firms clearly indicate their association with their principal companies when they export their products.

Very recently a mission representing the Japanese Hand and Small Tools Association and assisted by JETRO (Japanese External Trade Organization) visited Sialkot City. The mission was looking for potential suppliers of various tools to be distributed under Japanese trade marks. The mission was very much impressed by the workers' skill in using primary production tools. The result of this visit is still to be seen. While this report is being written, another mission sponsored by Japanese firms like Mitsui is in this country exploring fields where Japanese industrialists could join hands with Pakistanis in fabrication and export of goods, e.g., to the Middle East and East Africa, under Japanese trade marks.

Both the government and the progressive industrialists are sure of the future of sub-contracting, particularly that of international sub-contracting, as they are playing their part to the best of their ability for its development.

Now let us discuss the present extent of international sub-contracting industry-wise.

Transport equipment

Automobiles are assembled under licensing agreements with foreign firms. Presently there are four assemblers, who collectively employ nearly 1,000 workers. The foreign car-manufacturers supply components up to approximately 71 per cent of the total vehicle cost in CKD condition, the remaining 29 per cent of the components are now locally produced either at the assemblers' own factories or by their sub-contractors. Nearly 70 to 80 sub-contractors employing totally over 5,000 workers have been developed with the active assistance from their respective parent firms.

Motorcycles and scooters are also assembled by four units, employing nearly 1,000 workers. On the average, approximately 60 per cent of the total value of the components is supplied by the foreign principal firms in CKD condition, while the rest is produced in Pakistan. About 20 per cent of local contents are procured from over 30 sub-contractors. However, no international sub-contracting is involved.

Electric and electronic appliances

There are over a dozen firms in this industry. The most important among them are subsidiaries of foreign firms such as Johnson and Philips, Siemens, and Telefunken. Though these firms are primarily catering to the local market, they do export part of their production. Their total combined output is US$ 30 to 40 million.

There exists some sub-contracting among them and for them. The subsidiaries of foreign firms not only use the technical data provided by their principal firms but also send their products periodically to them for tests.

Almost all the electronic components for the radio and T.V. sets are imported, mostly from or through the principal firms and technical collaborators. Only some discrete electrical components and plastic, wood and metal parts are locally produced to the specifications and under guidance provided by the principal companies. Four firms, namely S.N.H. Industries; Audio Tech.; Pacific Traders and Associated Electronic, are unique in being entirely dependent on the ancillary industry for the supply of components and parts. Details of one of these units is given in the Case Studies section.

Attempts are being made to export radios and T.V. sets to the Middle East and East Africa. A large number of the products will, when exported, carry the principal firms' trade marks.

Textile and garments

Textile industry of Pakistan is relatively well-developed. Some firms have taken to open-ended spinning, which is the latest innovation in the industry, while double knit fabrics have been produced for a number of years. Both yarn and cloth have been exported for the last two decades or so. Recent additions are garments of all types including underwears.

In this industry, goods produced to own specifications are exported extensively, but a certain percentage of exports are fabricated to the specified requirements of the parent firms abroad. There are instances where both cloth and yarn are exported either with some foreign brands or without any brand on them to be marketed in a third country by the parent firm after imprinting their own brands. The execution of such orders sometimes help local producers to acquire new ideas and knowhow, as has been the case in blending of yarn, in weaving of fabrics for various uses, and in cutting and designing.

Sports goods

This is an industry which has existed in the border city of Sialkot in Punjab for well over a century. The range of its products covers almost all the outdoor games and some indoor games as well. There are 600 to 700 establishments, a few of which are medium and small in size, while the rest of them are cottage industries. These small units feed the exporters. While the larger firms procure up to about 25 per cent of their total sales from the cottage industries, the smaller exporters at times sub-contract up to 80 per cent of their requirements to them. Often, advance payments are made and raw materials are supplied to the sub-contractors.

One or two of the medium-sized firms have their own representatives in one or two foreign countries, to which their products are sent both for sale in the country concerned and for re-export to other countries. There are at least two firms which supply their products to foreign buyers according to the buyers' designs, specifications and requirements. For instance, Dunlop and Slazingers purchase rackets and other products from Ali Trading Co. of Sialkot for distribution in various countries under their own brands. Similarly, Greys of Cambridge supply cricket balls made in collaboration with and to the specifications of a U.K. firm. In one case, a parent company supplied cut segments of footballs which were re-exported after being stitched and finished. The order was fairly large and was further sub-contracted. Such business experience has taught the local fabricators about taste of consumers in foreign countries.

Printing

Printing industry is the latest entrant to the international sub-contracting. Enquiries had been made by certain parent companies, over the last few years, with a view to finding out if Pakistan could print some of their materials, particularly those for advertising in the Middle East and East Africa. Because the quality of printing was not up to the standard, none of these enquiries materialised. Recently, however, some of the firms imported computerised printing machinery, both at the suggestion of the parent firm and also at their own initiative. This has raised the quality of printing to the international standard, and foreign firms are now having their advertisement materials printed in Pakistan.

OBSTACLES TO INTERNATIONAL SUB-CONTRACTING

The limited spread of sub-contracting in the manufacture of parts and components in Pakistan can be attributed to a number of factors. First, small sub-contractors cannot easily get raw materials of desired specifications. Second, their production methods are not up to the standard. Third, they have not been able to convince the contract placing agency of their ability to handle the work independently.

Non-availability of basic raw materials is a particularly serious problem. Metals and chemicals are generally not produced in the country. As they have to be imported, they are not always available in the desired quality and quantity. Another problem is that individual sub-contractors' requirements of certain materials are often too small to attract foreign suppliers. Frequently, efforts are made to produce nearest possible substitutes to overcome the problem, but they are very reluctantly accepted even for trial purposes by the foreign parent firms, which are certainly quality-conscious. Interviewed parent firms also pointed out that small sub-contractors do not possess necessary precision machinery nor quality control equipment. On the other hand, it is obvious that the main advantage in securing a foreign parent firm is that parent firms supply not only technical knowhow but also in some cases, necessary raw materials.

GOVERNMENT'S POLICY

To boost export and international sub-contracting, the Government of Pakistan has instituted or plans to introduce certain schemes to aid the manufacturers.

The "Bonded Area" Scheme

Under this Scheme, an entire factory is cordoned off and segregated by the customs authorities. In this area, raw materials are imported without paying customs duty but against an indemnity bond. The products are directly exported from there and the indemnity bond is thereafter released. Exemption from import duty and sales tax reduces capital outlay and consequently production costs. There are some factories, particularly in the garment sector, which have been able to fulfil the required conditions and have been declared to be "bonded" under this scheme.

Industrial Free Zone Scheme

Another scheme, which is under serious consideration at present, is that of an Industrial Free Zone, where machinery and raw materials will be allowed to be imported by the entrepreneur without paying customs duty and sales tax. The products will be permitted to be exported without intervention of the customs authorities. They will, however, see to it that neither the raw materials nor the finished products infiltrate into local market outside the Free Zone area without authorisation and without paying customs duty and sales tax.

The Sub-contracting Exchange

Other proposals under consideration include the supply of complete information on the existing industry in Pakistan and its capability to the potential parent firms abroad. For this purpose a "Sub-contracting Exchange" is being set up with active assistance from the United Nations Industrial Development Organization (UNIDO). This Exchange would be a clearing house for detailed information on individual manufacturers in the country and would help the parent firms in locating potential sub-contractors.

RECOMMENDATIONS

1. Creation of a Sub-contracting Exchange and the Free Zone may be expedited. Implementation of such projects could bring parent firms in foreign countries into direct contact with sub-contractors in Pakistan and thus speed up the process of transfer of technology.
2. To the extent possible, a raw-material bank may be created to meet the needs of small-scale sub-contractors and those who cannot afford to carry a stock of raw materials and tools for their daily use. In this connection, a detailed survey may be carried out to assess the exact requirements in respect of both the kinds of raw materials needed and the quantities required.
3. The State enterprises, in particular those that have technical collabora-

tion with foreign firms, should encourage sub-contracting on the national level, as this would train the ancillary industry and small producers for work, connected with international sub-contracting. This would not only help to spread already existing knowhow but also to lay a foundation for acquiring further know-how both from foreign parent firms in Pakistan and their principals abroad.

4. Testing laboratory and other inspection facilities may be set up at all vulnerable centres, so that effective assistance in production techniques and quality control can be provided, particularly for very small sub-contractors.
5. After a Sub-contracting Exchange and a Free Zone have been created, very wide publicity should be given by both the Pakistani Missions abroad and international organisations concerned to introduce this country as a source of supply to foreign parent companies.
6. A corporation, which could import all the required raw materials and tools and could sell them to the smallest fabricators at no-loss-no-profit basis and on a day to a week's credit, is desired by the small fabricators. They in fact feel this corporation could act as their trading house vis-a-vis the foreign parent firms, and wish that the value of their semi-finished products be paid to them each evening.

SECTION 2: CASE STUDIES

(a) National Motors Limited

National Motors is one of the nationalised units of industry working under the Automobile Corporation of Pakistan Limited. The firm has been engaged in assembly-cum-manufacture of Bedford truck and bus chassis, load and bus bodies. Vauxhall Victor cars, pick-ups, vans and jeeps have also been assembled from time to time.

The firm is assembling and manufacturing vehicles under a licencing agreement with Vauxhall Motors Ltd., U.K., which provides necessary technical knowhow of assembly and parts-manufacturing. For this purpose, detailed drawings of components and tooling are made available. The locally fabricated components are tested by Vauxhall, which also provides facilities for training of the local firms' engineers, and trains them in the U.K. The British firm also assists in the procurement of various parts and components, which are manufactured in the U.K. and other countries.

National Motors is responsible for local manufacture of Bedford trucks excluding engine. Local contents are expected to go up to 50 per cent of material costs by December, 1976. With the existing production facilities it will be possible to achieve a local content of 67 per cent. For local procurement beyond this percentage, additional facilities will have to be created for items like front axle, propeller shaft, steering gear, servo brake and wheel with assistance from Vauxhall.

To achieve the indigenous content of 67 per cent, an extensive survey of the existing production facilities in the country was carried out by the engineers who, with technical advice from Vauxhall, have been able to select certain sub-contractors. At present, National Motors has over 60 sub-contractors and suppliers in Pakistan, who locally manufacture components of quality approved by Vauxhall for the Bedford trucks. The parent firm, with assistance from Vauxhall, has been providing various types of assistance needed by the sub-contractors, such as supply of technical know-how, personal visits by technical advisers, substantial advance payments, procurement of raw materials, and supply of dies and tooling. However, some difficulty is at times faced by the ancillary industry in the procurement of adequate raw materials. On such occasions, Vauxhall extends assistance.

(b) Johnson & Philips (Pakistan) Ltd.

This firm is a public limited concern established in 1961 in collaboration with the principal firm of the same name in Britain. The foreign partner owns 60 per cent of the equity amounting to US$655,000. Its main line of production is 11 K.V. switchgear, H.V. switchgear, EHV control panel and distribution appliances. Its total annual turnover is around US$3 million, approximately 20 per cent of which is exported to the Ministry of Electricity and Water, State Electricity Department, Qatar, etc.

The firm acts as a parent firm to many companies in the country, and as a sub-contractor to Siemens (Pak) Limited, to which it now supplies OCBs.

The principal firm provides various types of technical assistance: supply of detailed drawings of the components and tooling, and engineering data on the product testing methods, training of engineers and technicians whenever a new product is introduced, provision of advisers on the workshop practice, and organisation of refresher courses on the work floor, access to all its patents, etc.

On the other hand, Johnson and Philips helps its main sub-contractors. A good example is that of a sub-contractor who supplies ceramic components. Through the guidance provided, the quality of his products has improved over the time and he is now able to supply even quality products.

The main merits of international sub-contracting to the firm are to increase sales, utilise the available manufacturing capacity more fully, introduce goods to newer markets and new products to the existing markets, acquire technical knowhow through the principal firm, etc. The firm is making concerted efforts to increase its exports, which are normally made through quoting against tenders, and personal contacts with potential buyers. At times, enquiries are redirected by the principal firm. Products of the firm are all exported under the brand of the principal firm.

Technicians trained by the principal firm at times leave Johnson and Philips to join other firms. They take with them the knowledge and experience gained, which in turn help to improve the standard of the firm that they join.

(c) Siemens Pakistan Engineering Co., Ltd.

The firm is a limited concern which owns two factories: one was set up in 1955 and the other in 1962. As its name suggests, this is a subsidiary of Siemens of West Germany. It receives all its technical data and engineering assistance from the principal firm. Its engineering staff is trained at the principal firm's works, whenever necessary. The range of technical assistance from the principal firm is much the same as in the case of Johnson & Philips.

The firm manufactures electric motors, generators, transformers, switchgears and electro-medical equipment. In 1974-75, motors, transformers, switchgears and distribution boards were exported to the Philippines, Dubai, Abu Dhabi, etc. From 1977 onwards, it proposes to export goods worth US$400,000 a year. All the exported products carry the trade mark of Siemens.

(d) International Foundation & Garments (Pak) Ltd.

This manufacturer of ladies' underwears (brassieres only at present) went into production in 1971, and now has 246 employees. 50 per cent of its capital is owned by a foreign partner from Hong Kong, Triumph International.

Almost all its products are exported. To meet its raw material requirements, the firm imports approximately US$2.7 million worth of goods and purchases about US$3.1 million worth material in the local market. Because of bonded warehouse facilities the firm has no difficulty in importing raw materials and exporting its products.

The firm is a parent firm and a sub-contractor at the same time; it has a number of local sub-contractors while its exports carry the trade mark of Triumph International.

(e) Ali Trading Company Limited

This unit, a private limited enterprise, was established in 1956. Its main line of production is sports goods. Employing 272 persons, its annual production capacity is valued at US$1.5 million, of which US$1.3 million worth items can be exported. The firm's present production is, however, of the order of US$1.2 million all of which is sold abroad. It supplies goods to Dunlop and Slazingers, the two world renowned firms, whose offices in Germany, Switzerland, the United States, etc. place orders.

Part of the received orders are sub-contracted to smaller firms. Some of the cottage scale units produce sports goods of a really good quality. If they were trained and made quality conscious, they could produce goods of internationally accepted quality, and such training was felt essential by the parent firm, particularly because foreign markets are now becoming more and more competitive and quality-conscious.

Appendix I

List of Participants of the Study Meeting

Chief Expert

ILO	Mr. Susumu WATANABE Technology and Employment Branch Employment and Development Dept. International Labour Office CH 1211, Geneva 22 (P.O. Box 500) Switzerland

Country Experts

HONG KONG	Mr. Louis T.P. LAW Research Officer Hong Kong Productivity Centre 20th and 21st Floors Sincere Building 173 Des Voeux Road, Central Hong Kong
INDIA	Dr. S.K. SUBRAMANIAN Director (Technical Management) Management Development Institute A-21, Palam Marg Vasant Vihar New Delhi-110057
PAKISTAN	Mr. K.M. RIAZUDDIN Director (Engg-II) Department of Investment Promotion and Supplies Government of Pakistan Karachi

PHILIPPINES	Mr. Leopoldo S. TITULAR Project Officer Industrial Development Program Development Academy of the Philippines Parklane Building, Gambon St. Legaspi Village Makati, Metro Manila (P.O. Box 5160, Makati, Rizal)
SINGAPORE	Mr. LIU Shih Shin Investment Officer Economic Development Board 2nd Floor, Fullerton Bldg. P.O. Box 2692 Singapore 1

CAMS Representatives

Prof. Shigeru ISHIKAWA
Institute of Economic Research
Hitotsubashi University
Kunitachi-shi

Prof. Kounosuke ODAKA
Institute of Economic Research
Hitotsubashi University
Kunitachi-shi

Prof. Keinosuke ONO
Graduate School of Business Administration
Keio University
Yokohama-shi

Observers

Mr. Osamu OHKAWARA
Head
Overseas Technical Cooperation Division
Vocational Training Bureau
Ministry of Labour
Government of Japan
Tokyo

Mr. Hideo FUJIMORI
Economic Growth Division
Institute of Developing Economies
Tokyo

Appendix II

Recommendations of the Study Meeting

- Every effort should be made to overcome the problems of shortage and poor quality of raw materials which are among the major constraints on the development of both national and international sub-contracting.
- The quality-consciousness among local industrialists should be stimulated. National Productivity Organisations (NPOs) of the APO member countries and the APO should play a key role in this task by organising seminars, lectures and study tours, and by providing consultancy and information services.
- A high degree of standardisation of parts and components is much more difficult to attain in international sub-contracting than in national sub-contracting, since different parent firms in different countries have different specifications. Nevertheless, this may be achieved by either choosing parent firm and try to get as much work from it as possible.In certain industries the number of models should be reduced.
- As it is often technically impossible to produce internationally competitive products, in regard to price or quality or both, by old fashioned techniques, efforts should be made to overcome this problem by adopting different criteria of choice of technology according to the market for which the product is intended or according to the production process.
- In order to encourage modernisation of capital equipment and bring up the technological standard of sub-contractors and would-be sub-contractors to the international level, financial assistance from the local government might be necessary.
- Legislation aiming at ensuring the parent firms' adherence to contractual clauses preventing their abuse of the sub-contractors may be considered by member countries in line with the Anti-Monopoly Law and the Payment of Sub-contractors (Prevention of Delay) Law in Japan.
- Industrial associations, or sub-contractors' consortia, may be encouraged to promote co-operation among sub-contractors in individual industries,

in regard to e.g., quality control and procurement of raw materials. This may also help strengthen sub-contractors' bargaining position vis-à-vis their parent firms, or at a time of business recession.

- Preferential treatment in regard to taxation or loans may be granted to those parent firms which cooperate in developing sub-contractors in certain fields of activities.
- Cultural exchange may merit special encouragement by governments to pave the way for promotion of international sub-contracting and related technology transfer.